THE MYTH OF ARTIFICIAL INTELLIGENCE

THE MYTH OF ARTIFICIAL INTELLIGENCE

. . .

Why Computers Can't Think the Way We Do

. . .

ERIK J. LARSON

The Belknap Press of Harvard University Press
Cambridge, Massachusetts · London, England

Printed in the United States of America

First Harvard University Press paperback edition, 2022
Third printing

Library of Congress Cataloging-in-Publication Data

Names: Larson, Erik J. (Erik John), author.
Title: The myth of artificial intelligence : why computers can't think the way we do / Erik J. Larson.
Description: Cambridge, Massachusetts : The Belknap Press of Harvard University Press, 2021. | Includes bibliographical references and index.
Identifiers: LCCN 2020050249 | ISBN 9780674983519 (cloth) | ISBN 9780674278660 (pbk.)
Subjects: LCSH: Artificial intelligence. | Intellect. | Inference. | Logic. | Natural language processing (Computer science) | Neurosciences.
Classification: LCC Q335 .L37 2021 | DDC 006.3—dc23
LC record available at https://lccn.loc.gov/2020050249

To Brooke and Ben

CONTENTS

INTRODUCTION

In the pages of this book you will read about the myth of artificial intelligence. The myth is not that true AI is possible. As to that, the future of AI is a scientific unknown. The myth of artificial intelligence is that its arrival is inevitable, and only a matter of time—that we have already embarked on the path that will lead to human-level AI, and then superintelligence. We have not. The path exists only in our imaginations. Yet the inevitability of AI is so ingrained in popular discussion—promoted by media pundits, thought leaders like Elon Musk, and even many AI scientists (though certainly not all)—that arguing against it is often taken as a form of Luddism, or at the very least a shortsighted view of the future of technology and a dangerous failure to prepare for a world of intelligent machines.

As I will show, the science of AI has uncovered a very large mystery at the heart of intelligence, which no one currently has a clue how to solve. Proponents of AI have huge incentives to minimize its known limitations. After all, AI is big business, and it's increasingly dominant in culture. Yet the possibilities for future AI systems are limited by what we currently know about the nature of intelligence, whether we like it or not. And here we should say it directly: all evidence suggests that human and machine intelligence are radically different. The myth of AI insists that the differences are only temporary, and that more powerful systems will eventually erase them. Futurists like

Ray Kurzweil and philosopher Nick Bostrom, prominent purveyors of the myth, talk not only as if human-level AI were inevitable, but as if, soon after its arrival, superintelligent machines would leave us far behind.

This book explains two important aspects of the AI myth, one scientific and one cultural. The scientific part of the myth assumes that we need only keep "chipping away" at the challenge of general intelligence by making progress on narrow feats of intelligence, like playing games or recognizing images. This is a profound mistake: success on narrow applications gets us not one step closer to general intelligence. The inferences that systems require for general intelligence—to read a newspaper, or hold a basic conversation, or become a helpmeet like Rosie the Robot in *The Jetsons*—cannot be programmed, learned, or engineered with our current knowledge of AI. As we successfully apply simpler, narrow versions of intelligence that benefit from faster computers and lots of data, we are not making incremental progress, but rather picking low-hanging fruit. The jump to general "common sense" is completely different, and there's no known path from the one to the other. No algorithm exists for general intelligence. And we have good reason to be skeptical that such an algorithm will emerge through further efforts on deep learning systems or any other approach popular today. Much more likely, it will require a major scientific breakthrough, and no one currently has the slightest idea what such a breakthrough would even look like, let alone the details of getting to it.

Mythology about AI is bad, then, because it covers up a scientific mystery in endless talk of ongoing progress. The myth props up belief in inevitable success, but genuine respect for science should bring us back to the drawing board. This brings us to the second subject of these pages: the cultural consequences of the myth. Pursuing the myth is not a good way to follow "the smart money," or even a neutral stance. It is bad for science, and it is bad for us. Why? One reason is that we are unlikely to get innovation if we choose to ignore a core mystery rather than face up to it. A healthy culture for innovation emphasizes exploring unknowns, not hyping extensions of existing methods—especially when these methods have been shown to be inadequate to take us much further. Mythology about inevitable success in AI tends to extinguish the very culture of invention necessary for real progress—with or without human-level AI. The myth also encourages resignation to the creep of a machine-land, where genuine invention is sidelined in favor of futuristic talk advocating current approaches, often from entrenched interests.

Who should read this book? Certainly, anyone should who is excited about AI but wonders why it is always ten or twenty years away. There is a scientific reason for this, which I explain. You should also read this book if you think AI's advance toward superintelligence is inevitable and worry about what to do when it arrives. While I cannot prove that AI overlords will not one day appear, I can give you reason to seriously discount the prospects of that scenario. Most generally, you should read this book if you are simply curious yet confused about the widespread hype surrounding AI in our society. I will explain the origins of the myth of AI, what we know and don't know about the prospects of actually achieving human-level AI, and why we need to better appreciate the only true intelligence we know—our own.

IN THIS BOOK

In Part One, The Simplified World, I explain how our AI culture has simplified ideas about people, while expanding ideas about technology. This began with AI's founder, Alan Turing, and involved understandable but unfortunate simplifications I call "intelligence errors." Initial errors were magnified into an ideology by Turing's friend and statistician, I. J. Good, who introduced the idea of "ultraintelligence" as the predictable result once human-level AI had been achieved.

Between Turing and Good, we see the modern myth of AI take shape. Its development has landed us in an era of what I call technological kitsch—cheap imitations of deeper ideas that cut off intelligent engagement and weaken our culture. Kitsch tells us how to think and how to feel. The purveyors of kitsch benefit, while the consumers of kitsch experience a loss. They—we—end up in a shallow world.

In Part Two, The Problem of Inference, I argue that the only type of inference—thinking, in other words—that will work for human-level AI (or anything even close to it) is the one we don't have a clue how to program or engineer. The problem of inference goes to the heart of the AI debate because it deals directly with intelligence, in people or machines. Our knowledge of the various types of inference dates back to Aristotle and other ancient Greeks, and has been developed in the fields of logic and mathematics. Inference is already described using formal, symbolic systems like computer programs, so a very clear view of the project of engineering intelligence can be gained by exploring inference. There are three types. Classic AI explored one (deduction), modern AI explores another (induction). The third type (abduction) makes for general intelligence, and, surprise, no one is working on it—at all.[1] Finally, since each type of inference is distinct—meaning, one type cannot be reduced to another—we know that failure to build AI systems using the type of inference undergirding general intelligence will result in failure to make progress toward artificial general intelligence, or AGI.

In Part Three, The Future of the Myth, I argue that the myth has very bad consequences if taken seriously, because it subverts science. In particular, it erodes a culture of human intelligence and invention, which is necessary for the very breakthroughs we will need to understand our own future. Data science (the application of AI to "big data") is at best a prosthetic for human ingenuity, which if used correctly can help us deal with our modern "data deluge." If used as a replacement for individual intelligence, it tends to chew up investment without delivering results. I explain, in particular, how the myth has negatively affected research in neuroscience, among other recent scientific pursuits. The price we are paying for the myth is too high. Since we have no good scientific reason to believe the myth is true, and every reason to reject it for the purpose of our own future flourishing, we need to radically rethink the discussion about AI.

Part I

THE SIMPLIFIED WORLD

Chapter 1

• • •

THE INTELLIGENCE ERROR

The story of artificial intelligence starts with the ideas of someone who had immense human intelligence: the computer pioneer Alan Turing.

In 1950 Turing published a provocative paper, "Computing Machinery and Intelligence," about the possibility of intelligent machines.[1] The paper was bold, coming at a time when computers were new and unimpressive by today's standards. Slow, heavy pieces of hardware sped up scientific calculations like code breaking. After much preparation, they could be fed physical equations and initial conditions and crank out the radius of a nuclear blast. IBM quickly grasped their potential for replacing humans doing calculations for businesses, like updating spreadsheets. But viewing computers as "thinking" took imagination.

Turing's proposal was based on a popular entertainment called the "imitation game." In the original game, a man and a woman are hidden from view. A third person, the interrogator, relays questions to one of them at a time and, by reading the answers, attempts to determine which is the man and which the woman. The twist is that the man has to try to deceive the interrogator while the woman tries to assist him—making replies from either side suspect. Turing replaced the man and woman with a computer and a human. Thus began what we now call the Turing test: a computer and a human receive typed questions from a human judge, and if the judge can't accurately identify which is the computer, the computer wins. Turing argued that with such an outcome, we have no good reason to define the machine as unintelligent, regardless of whether it is human or not. Thus, the question of whether a machine has intelligence replaces the question of whether it can truly think.

The Turing test is actually very difficult—no computer has ever passed it. Turing, of course, didn't know this long-term result in 1950; however, by replacing pesky philosophical questions about "consciousness" and "thinking" with a test of observable output, he encouraged the view of AI as a legitimate science with a well-defined aim. As AI took shape in the 1950s, many of its pioneers and supporters agreed with Turing: any computer holding a sustained and convincing conversation with a person would be, most of us would grant, doing something that requires thinking (whatever that is).

TURING'S INTUITION / INGENUITY DISTINCTION

Turing had made his reputation as a mathematician long before he began writing about AI. In 1936, he published a short mathematical paper on the precise meaning of "computer," which at the time referred to a person working through a sequence of steps to get a definite result (like performing a calculation).[2] In this paper, he replaced the human computer with the idea of a machine doing the same work. The paper ventured into difficult mathematics. But in its treatment of machines it made no reference to human thinking or the mind. Machines can run automatically, Turing said, and the problems they solve do not require any "external" help, or intelligence. This external intelligence—the human factor—is what mathematicians sometimes call "intuition."

Turing's 1936 work on computing machines helped launch computer science as a discipline and was an important contribution to mathematical logic. Still, Turing apparently thought that his early definition missed something essential. In fact, the same idea of the mind or human faculties assisting problem-solving appeared two years later in his PhD thesis, a clever but ultimately unsuccessful attempt to bypass a result from the Austrian-born mathematical logician Kurt Gödel (more on this later). Turing's thesis contains this curious passage about intuition, which he compares with another mental capability he calls ingenuity:

> Mathematical reasoning may be regarded rather schematically as the exercise of a combination of two faculties, which we may call intuition and ingenuity. The activity of the intuition consists in making spontaneous judgments which are not the result of conscious trains of reasoning. These judgments are often but by no means invariably correct (leaving aside the question as to what is meant by "correct"). Often it is possible to find some other way of verifying the correctness of an intuitive judgment. One may for instance judge that all positive integers are uniquely factorable into primes; a detailed mathematical argument leads to the same result. It will also involve intuitive judgments, but they will be ones less open to criticism than the original judgment about factorization. I shall not attempt to explain this idea of "intuition" any more explicitly.

Turing then moves on to explain ingenuity: "The exercise of ingenuity in mathematics consists in aiding the intuition through suitable arrangements of propositions, and perhaps geometrical figures or drawings. It is intended that when these are really well arranged the validity of the intuitive steps which are required cannot seriously be doubted."[3]

Though his language is framed for specialists, Turing is pointing out the obvious: mathematicians typically select problems or "see" an interesting problem to work on using some capacity that at least *seems* indivisible into steps—and therefore not obviously amenable to computer programming.

GÖDEL'S INSIGHT

Gödel, too, was thinking about mechanical intelligence. Like Turing, he was obsessed with the distinction between ingenuity (mechanics) and intuition (mind). His distinction was essentially the same as Turing's, in different language: proof versus truth (or "proof-theory" versus "model-theory" in mathematics lingo). Are the concepts of proof and truth, Gödel wondered, in the end the same? If so, mathematics and even science itself might be understood purely mechanically. Human thinking in this view would be mechanical, too. The concept of AI, though the term remained to be coined, hovered above the question. Is the mind's intuition, its ability to grasp truth and meaning, reducible to a machine, to computation?

This was Gödel's question. In answering it, he ran into a snag that would soon make him world-famous. In 1931, Gödel published two theorems of mathematical logic known as his incompleteness theorems. In them, he demonstrated the inherent limitations of all formal mathematical systems. It was a brilliant stroke. Gödel showed unmistakably that mathematics—*all* of mathematics, with certain straightforward assumptions—is, strictly speaking, not mechanical or "formalizable." More specifically, Gödel proved that there must exist some statements in any formal (mathematical or computational) system that are True, with capital-T standing, yet not provable in the system itself using any of its rules. The True statement can be recognized by a human mind, but is (provably) not provable by the system it's formulated in.

How did Gödel reach this conclusion? The details are complicated and technical, but Gödel's basic idea is that we can treat a mathematical system complicated enough to do addition as a system of meaning, almost like a natural language such as English or German—and the same applies to all more complicated systems. By treating it this way, we enable the system to talk about itself. It can say about itself, for instance, that it has certain limitations. This was Gödel's insight.

Formal systems like those in mathematics allow for the precise expression of truth and falsehood. Typically, we establish truth by using the tools of proof—we use rules to prove something, so we know it's definitely true. But are there true statements that can't be proven? Can the mind know things the system cannot? In the simple case of arithmetic, we express truths by writing equations like "$2 + 2 = 4$." Ordinary equations are true statements in the system of arithmetic, and they are provable using the rules of arithmetic. Here, provable equals true. Mathematicians before Gödel thought all of mathematics had this property. This implied that machines could crank out all truths in different mathematical systems by simply applying the rules correctly. It's a beautiful idea. It's just not true.

Gödel hit upon the rare but powerful property of self-reference. Mathematical versions of self-referring expressions, such as "This statement is not provable in this system," can be constructed without breaking the rules of mathematical systems. But the so-called self-referring "Gödel statements" introduce contradictions into mathematics: if they are true, then they are unprovable. If they are false, then because they say they are unprovable, they are actually true. True means false, and false means true—a contradiction.

Going back to the concept of intuition, we humans can see that the Gödel statement is in fact true, but because of Gödel's result, we also know that the system's rules can't prove it—the system is in effect blind to something not covered by its rules.[4] Truth and provability pull apart. Perhaps mind and machine do, as well. The purely formal system has limits, at any rate. It cannot prove in its own language something that is true. In other words, we can see something that the computer cannot.[5]

Gödel's result dealt a massive blow to a popular idea at the time, that all of mathematics could be converted into rule-based operations, cranking out mathematical truths one by one. The zeitgeist was formalism—not talk of minds, spirits, souls, and the like. The formalist movement in mathematics signaled a broader turn by intellectuals toward scientific materialism, and in particular, logical positivism—a movement dedicated to eradicating traditional metaphysics like Platonism, with its abstract Forms that couldn't be observed with the senses, and traditional notions in religion like the existence of God. The world was turning to the idea of precision machines, in effect. And no one took up the formalist cause as vigorously as the German mathematician David Hilbert.

HILBERT'S CHALLENGE

At the outset of the twentieth century (before Gödel), David Hilbert had issued a challenge to the mathematical world: show that all of mathematics rested on a secure foundation. Hilbert's worry was understandable. If the purely formal rules of mathematics can't prove any and all truths, it's at least theoretically possible for mathematics to disguise contradictions and nonsense. A contradiction buried somewhere in mathematics ruins everything, because from a contradiction anything can be proven. Formalism then becomes useless.

Hilbert expressed the dream of all formalists, to prove finally that mathematics is a closed system governed only by rules. Truth is just "proof." We acquire knowledge by simply tracing the "code" of a proof and confirming no rules were violated. The larger dream, thinly disguised, was really a worldview, a picture of the universe as itself a mechanism. AI began taking shape as an idea, a philosophical position that might also be proven. Formalism treated intelligence as a rule-based process. A machine.

Hilbert issued his challenge at the Second International Congress of Mathematicians in Paris in 1900. The intellectual world was listening. His challenge had three main parts: to prove that mathematics was complete; to prove that mathematics was consistent; and to prove that mathematics was decidable.

Gödel dealt the first and second parts of Hilbert's challenge a death blow with the publication of his incompleteness theorems in 1931. The question of decidability was left unanswered. A system is decidable if there is a definite procedure (a proof, or sequence of deterministic, obvious steps) to establish whether any statement constructed using the rules of the system is true or false. The statement 2 + 2=4 must be True, and 2 + 2=5 must be False. And so for all statements that one can validly make using the symbols and rules of the system. Since arithmetic was thought to be the foundation of mathematics, proving mathematics was decidable amounted to proving the result for arithmetic and its extensions. This would amount to saying that mathematicians, playing a "game" with rules and symbols (the formalist idea), were in fact playing a valid game that never led to contradiction or absurdity.

Turing was fascinated with Gödel's result, which demonstrated not the power of formal systems but rather their limitations. He took up work on the remaining part of Hilbert's challenge, and began thinking in earnest about whether a decision procedure for formal systems might exist. By 1936, in his paper "Computable Numbers," he proved that it must not. Turing realized that Gödel's use of self-reference also applied to questions about decision procedures or, in effect, computer programs. In particular, he realized that there must exist (real) numbers that *no* definite method could "calculate," by writing out their decimal expansion, digit by digit. He imported a result from the nineteenth-century mathematician Georg Cantor, who proved that real numbers (those with a decimal expansion) were more numerous than the integers, even though real numbers and integers are both infinite. Turing stood on the shoulders of giants, perhaps. But in the end, his work in "Computable Numbers" proved again a negative. It was a limiting result: no universal decision procedure was possible. In other words, rules—even in mathematics—aren't enough. Hilbert was wrong.[6]

IMPLICATIONS FOR AI

What is important to AI here is this: Turing disproved that mathematics was decidable by inventing a machine, a deterministic machine, requiring no insight or intelligence to solve problems. Today, we refer to his abstract formulation of a machine as a Turing machine. I am typing on one right now. Turing machines are computers. It is one of the great ironies of intellectual history that the theoretical framework for computation was put in place as a side-thought, a means to another end. While working to disprove that mathematics itself was decidable, Turing first invented something precise and mechanical, the computer.

In his 1938 PhD thesis, Turing hoped that formal systems might be extended by including additional rules (then sets of rules, and sets of sets of rules) that could handle the "Gödel problem." He discovered, rather, that the new, more powerful system would have a new, more complicated Gödel problem. There was no way around Gödel's incompleteness. Buried in the complexities of Turing's discussion of formal systems, however, is an odd suggestion, relevant to the possibility of AI. Perhaps the faculty of intuition cannot be reduced to an algorithm, to the rules of a system?

Turing wanted to find a way out of Gödel's limiting result in his 1938 thesis, but he discovered that this was impossible. Instead, he switched gears, exploring how, as he put it, to "greatly reduce" the requirement of human intuition when doing calculations. His thesis considered the powers of ingenuity, by creating ever more complicated systems of rules. (Ingenuity, it turned out, could become universal—there are machines that can take as input other machines, and thus run all the machines that can be built. This insight, technically a universal Turing machine and not a simple Turing machine, was to become the digital computer.) But in his formal work on computing, Turing had (perhaps inadvertently) let the cat out of the bag. By allowing for intuition as distinct from and outside of the operations of a purely formal system like a computer, Turing in effect suggested that there may be differences between computer programs that do math and mathematicians.

It was a curious turn, therefore, that Turing made from his early work in the 1930s to the more wide-ranging speculation about the possibility of intelligent computers in "Computing Machinery and Intelligence," published a little over a decade later. By 1950, discussion of intuition disappeared from Turing's writings about the implications of Gödel. His interests turned, in effect, to the possibility that computers might become "intuition-machines" themselves. In essence, he decided that Gödel's result didn't apply to the question of AI: if we humans are highly advanced computers, Gödel's result means only that there are some statements that we cannot understand or see to be true, just as with less complicated computers. The statements might be fantastically complicated and interesting. Or, possibly, they might be banal yet overwhelmingly complex. Gödel's result left open the question of whether minds were just very complicated machines, with very complicated limitations.

Intuition, in other words, had become part of Turing's ideas about machines and their powers. Gödel's result couldn't say (to Turing, anyway) whether minds were machines or not. On the one hand, incompleteness says that some statements can be seen to be true using intuition, but cannot be proved by a computer using ingenuity. On the other hand, a more powerful computer can use more axioms (or more bits of relevant code) and prove the result—thus showing that intuition is not beyond computation for that problem. This becomes an arms race: more and more powerful ingenuity substituting for intuition on more and more complicated problems. No one can say who wins the race, so no one can make a case—using the incompleteness result—about the inherent differences between intuition (mind) and ingenuity (machine). But as Turing no doubt knew, if this were true, then so too was at least the possibility of artificial intelligence.

Thus, between 1938 and 1950, Turing had a change of heart about ingenuity and intuition. In 1938, intuition was the mysterious "power of selection" that helped mathematicians decide which systems to work with and what problems to solve. Intuition was not something in the computer. It was something that decided things about the computer. In 1938, Turing thought intuition wasn't part of any system, which suggested not only that minds and machines were fundamentally different but that AI-as-human-thinking was well-nigh impossible.

Yet by 1950 he had reversed his position. With the Turing test, he offered a challenge for skeptics and a sort of defense of intuition in machines, asking in effect: Why not? This was a radical about-face. A new view of intelligence, it seemed, was taking shape.

Why the shift? Something outside the world of strict mathematics and logic and formal systems had happened to Turing between 1938 and 1950. It had happened, in fact, to all of Great Britain, and indeed to most of the world. What happened was the Second World War.

Chapter 2

. . .

TURING AT BLETCHLEY

The game of chess fascinated Turing—as it did his wartime colleague, mathematician I. J. "Jack" Good. The two would play against each other (Good usually won) and work out decision procedures and rules of thumb for winning moves. Playing chess involves following the rules of the game (ingenuity), and it also seems to require insight (intuition) into which rules to choose given different positions on the game board. To win at chess, it is not enough to apply the rules; you have to know which rules to select in the first place.

Turing saw chess as a handy (and no doubt entertaining) way to think about machines and the possibility of giving them intuition. Across the Atlantic, the founder of modern information theory, Turing's colleague and friend Claude Shannon at Bell Labs, was also thinking about chess. He later built one of the first chess-playing computers, an extension of work he had done earlier on a proto-computer called the "differential analyzer," which could convert certain problems in calculus into mechanical procedures.[1]

THE SIMPLIFICATION OF INTELLIGENCE BEGINS

Chess fascinated Turing and his colleagues in part because it seemed that a computer could be programmed to play it, without the human programmer needing to know everything in advance. Because computing devices implemented logical operators like IF-THEN, OR, and AND, a program (set of instructions) could be run, and it could produce different results depending on the scenarios it encountered while running through the instructions. This ability to change course depending on what it "saw" seemed to Turing and his colleagues to simulate a core aspect of human thinking.[2]

The chess players—Turing, Good, Shannon, and others—were also thinking about another mathematical problem whose stakes were much higher. They were working for their governments, helping to crack secret codes used by Germany to coordinate attacks on commercial and military ships crossing the English Channel and the Atlantic. Turing found himself engaged in a desperate effort to help defeat Nazi Germany in the Second World War, and it was his ideas about computation that helped turn the tide of war.

BLETCHLEY PARK

Bletchley Park, situated unobtrusively in a small town out of the path of bombs falling in London and metropolitan Britain, was a research facility set up to help uncover the locations of German U-boats—submarines—which were laying waste to shipping routes in the English Channel. U-boats were a major problem for the Allied forces, sinking thousands of ships and destroying huge quantities of supplies and equipment. To maintain the war effort, Britain required imports amounting to thirty million tons a year. The U-boats were at one point depleting this by 200,000 tons a month, a significant, potentially catastrophic, and for a time largely unanswered German strategy in the war. In response, the British government assembled a group of talented cryptanalysts, chess players, and mathematicians to investigate how to crack U-boat communications, known as ciphers. (A cipher is a disguised message. To decipher a message is to convert it back to readable text).[3]

The codes were generated by a typewriter-looking device known as the Enigma, a kind of machine that had been in commercial use since the 1920s but that the Germans had strengthened significantly for use in the war. Modified Enigmas were used for all strategic communications in the Nazi war effort. The Luftwaffe, for instance, used the Enigma machine in its conduct of the air war, as did the Kriegsmarine in its naval operations. Messages encrypted with the modified Enigma were widely thought to be undecipherable.

Turing's role in Bletchley and his subsequent rise to national hero after the war is a story that has been told many times. (In 2014, the major motion picture *The Imitation Game* dramatized his work at Bletchley, as well as his subsequent role in developing computers.) Turing's major breakthrough was, by pure mathematical standards, relatively uninteresting because it exploited an old idea from deductive logic. The method that he and others half-jokingly referred to as "Turingismus" involved eliminating large numbers of possible solutions to Enigma codes by finding combinations with contradictions. Contradictory combinations are impossible; we cannot have both "A" and "not-A" in some logical system, just as we cannot be both "at the store" and "at home" at the same time. Turingismus was a winning idea, and became a huge success at Bletchley. It did what was required of the "boy geniuses" sequestered in the think tank by speeding up the task of decrypting Enigma messages. Other scientists devised different strategies for cracking the codes at Bletchley.[4] Ideas were tested on a machine called a Bombe—its tongue-in-cheek name borrowed from a predecessor machine in Poland, the Bomba, and possibly inspired by the small noise made when a calculation was finished. Think of the Bombe as a proto-computer, capable of running different programs.

The advantage in war swung from the Axis to Allied powers by 1943 or thereabouts, in no small part because of the sustained effort of the Bletchley code-crackers. The team was a celebrated success, and its members became war heroes. Careers were made. Bletchley, meanwhile, also proved a haven for thinking about computation: Bombes were machines, and they ran programs to solve problems that humans, by themselves, could not.

INTUITIVE MACHINES? NO.

For Turing, Bletchley played a major role in crystallizing his ideas about the possibility of intelligent machines. Like his colleagues Jack Good and Claude Shannon, Turing saw the power and utility of their "brain games" as cryptanalysts during the war: they could decipher messages that were otherwise completely opaque to the military. The new methods of computation were not just interesting for considering automated chess-playing. Computation could, quite literally, sink warships.

Turing was thinking about an abstraction (yet again): minds and machines, or the general idea of intelligence. But there was something odd about his view of what it meant. In the 1940s, intelligence was a trait not typically attributed to formal systems like the purely mechanical code-breaking Bombes of Bletchley. Gödel had demonstrated that, in general, truth cannot be reduced to formality, as in playing a formal game with a set sequence of rules—but recall that his proof left open the question of whether specific machines might actually incorporate the intuition that minds use to make choices about rules to follow, even if no supreme system could exist that could prove everything (which Gödel had shown so definitely in 1931).

After Bletchley, Turing turned increasingly to the question of whether powerful machines could be built that used intuition and ingenuity. The vast number of possible combinations to check to decipher German codes swamped human intuition. But systems with the right programs could accomplish the task by simplifying such vast mathematical possibilities. To Turing, this suggested that intuition could be embodied in machines. In other words, the success at Bletchley implied that perhaps an artificial intelligence could be built.

To make sense of his line of thought, however, some particular idea about "intelligence" had to be settled on. Intelligence as displayed by humans needed to be reducible—analyzable—in terms of the powers of a machine. In essence, intelligence had to be reducible to problem-solving. That is what playing chess is, after all, and that is what breaking a code is, as well.

And here we have it: Turing's great genius, and his great error, was in thinking that human intelligence reduces to problem-solving. Whether or not the ideas about intelligent machines in his 1950 "Computing Machinery and Intelligence" became explicit in the war years, it is clear that his experience at Bletchley crystallized his later view of AI, and it is clear that AI in turn followed closely and without necessary self-analysis precisely in his path.

But a closer look at the Bletchley code-cracking success immediately reveals a dangerous simplification in the philosophical ideas about man and machine. Bletchley was an intelligent system—a coordination of military efforts (including spying and espionage, as well as capture of enemy vessels), social intelligence between the military and the various scientists and engineers at Bletchley, and (as with all of life) sometimes sheer dumb luck. In truth, as a practical reality, the German-modified Enigma was unbreakable by purely mechanical means. The Germans knew this based on mathematical arguments about the difficulty of mechanical deciphering. Part of Bletchley's success was, ironically, the stubborn confidence of Nazi commanders in the impregnability of the Enigma ciphers—thus they failed at crucial times to modify or strengthen the machines after discovering certain ciphers had been cracked, blaming covert spying operations rather than scientific defeat. But the fog of war mixes together not just new technologies but new forms of human and social intelligence. War is not chess.

Early in the war, for instance, Polish forces had recovered important fragments of Enigma communications that later provided invaluable clues to Bletchley efforts. The Poles had used these fragments (along with others from Russian sources) to develop their own, simpler Bomba as early as 1938. Turing's much improved version in the early months of 1940—the Bombe using his "Turingismus"—relied on the early work the Poles made possible by events on the battlefield. Turing, too, would see his own design improved in response to improvements in the Enigma by his colleague Gordon Welchman, by which a "diagonal board" was added to further simplify the search for contradictions.[5] Here were two human minds, using intuition, working together socially.

More events in the theater of war proved vitally important. Off the shores of Norway, a British aircraft carrier was sunk on June 8, 1940. The attack provided the location of German U-boats, albeit at the heavy cost of many sailors left at the bottom of the sea. Just weeks before, in late April 1940, the German patrol boat *VP2623*, a particularly devastating member of the fleet, was captured with a trove of Enigma evidence inside. The necessary pieces of the Enigma puzzle were getting into Allied hands, and finding their way to the Bletchley group.

These bits and pieces by themselves were grossly inadequate for quick deciphering of future German communications, amounting to what one Turing biographer called "guesswork" for Bletchley cryptanalysts. But they facilitated an all-important first step in figuring out how to program the Bombes. Turing and colleagues called it the "weight of evidence," borrowing a term coined by the American scientist and logician C. S. Peirce (who is prominently featured in Part Two of this book).[6]

Weight of evidence can be understood by mathematicians in different ways, but for Bletchley's success (and for larger issues regarding AI) it amounts to the application of informed guesses, or intuition, to give direction to ingenuity, or machines. A scrap of deciphered text recovered from a captured U-boat could mean anything, just as a white ball found near a bag of white balls could mean anything, but in each case, we can make intelligent guesses to understand what happened. We think the ball is very likely from the bag, even though we didn't see it taken out. Still, it's a guess. Guesses of this sort can't be proven true, but the better human intuition does at setting initial conditions for devising mechanical procedures, the better chance those procedures have of terminating on desired outcomes, rather than, say, running on aimlessly in false or misleading directions. Weight of evidence—guessing—made Bombes work.

Bletchley scientists were not merely feeding information into Bombes, leaving them to do the tireless and important work of eliminating millions of incorrect codes or ciphers. To be sure, the Bombes were necessary—this is what Turing saw so clearly, and what no doubt suffused his imagination with the possibility that his "mechanical procedures" could reproduce or supersede human intelligence. But the fact is that the Bletchley group was first and foremost engaged in guesswork. They were forming hypotheses by recognizing the clues hidden in the patchwork of scraps of instructions, ciphers, and messages coming in from the battlefield. Guessing is known in science as forming hypotheses (a term Charles Sanders Peirce also used), and it is absolutely fundamental to the advancement of human knowledge. Small wonder then that the Bletchley effort amounted to a system of guessing well. Its sine qua non was not mechanical but rather what we might call initial intelligent observation. The Bombes had to be pointed at something, and then set on their course.

In line with a theme we will explore in Part Two, Peirce had recognized early on, by the late nineteenth century, that every observation that shapes the complex ideas and judgments of intelligence begins with a guess, or what he called an abduction:

> Looking out of my window this lovely spring morning I see an azalea in full bloom. No, no! I do not see that; though that is the only way I can describe what I see. That is a proposition, a sentence, a fact; but what I perceive is not proposition, sentence,

> fact, but only an image which I make intelligible in part by means of a statement of fact. This statement is abstract; but what I see is concrete. I perform an abduction when I [do so much] as express in a sentence anything I see. The truth is that the whole fabric of our knowledge is one matted felt of pure hypothesis confirmed and refined by induction. Not the smallest advance can be made in knowledge beyond the stage of vacant staring, without making an abduction at every step.[7]

Turing and his colleagues at Bletchley were winning a war that had turned from command and control to intelligence by making, in effect, intelligent abductions at every step. At some level, Turing no doubt understood this (recall the discussion of intuition in his 1938 thesis on ordinal numbers), but it seems not to have had an appreciable effect on his later ideas about the nature of intelligence and the possibility of intelligent machines. However brilliantly, he was formulating a simplification of real intelligence. He was getting rid of the concept that had so transfixed him earlier, of intuition. Of guessing.

ON SOCIAL INTELLIGENCE (AN IMPORTANT ASIDE)

Social intelligence is also conspicuously left out of Turing's puzzle-solving view of intelligence. This is of utmost importance for understanding the future development of AI. Turing, for instance, disliked viewing thinking or intelligence as something social or situational.[8] Yet the Bletchley success was in fact part of a vast system that extended far outside its cloistered walls. A massive effort was underway. It would soon pull in the United States and the work of scientists like Shannon at Bell Labs, as well as scientists at Princeton's celebrated Institute for Advanced Studies—where Einstein, Gödel, and John Von Neumann all had appointments. The expanded, human-machine system is actually much more realistic as a model of how actual real-world problems get solved—of which, world war must certainly count among the most complex and important.

AI's tone-deafness on social or situational intelligence has been noted before, more recently by machine learning scientist François Chollet, who summarizes his critique of Turing's (and, more broadly, the AI field's) view of intelligence nicely. First, intelligence is *situational*—there is no such thing as general intelligence. Your brain is one piece in a broader system which includes your body, your environment, other humans, and culture as a whole. Second, it is *contextual*—far from existing in a vacuum, any individual intelligence will always be both defined and limited by its environment. (And currently, the environment, not the brain, is acting as the bottleneck to intelligence.) Third, human intelligence is largely *externalized,* contained not in your brain but in your civilization. Think of individuals as tools, whose brains are modules in a cognitive system much larger than themselves—a system that is self-improving and has been for a long time.[9]

In Turing's language, intuition might be programmable into a machine, but Chollet and similar critics claim that it cannot reach the level of human intelligence. In fact, the idea of programming intuition ignores a fundamental fact about our own smarts. Humans have social intelligence. We have emotional intelligence. We use our minds other than to solve problems and puzzles, however complex (or rather, *especially* when the problems are complex).

Turing, the evidence suggests, decisively rejected this view of a person, instead coming to believe that all of human thought could be understood, in effect, as the "breaking" of "codes"—the solving of puzzles—and the playing of a game like chess. The important point is this: sometime in the 1940s, after his work at Bletchley but certainly by the time of his 1950 paper prefiguring AI, Turing had settled his thoughts on a simplified view of intelligence. This was an egregious error and, further, one that has been passed down through generations of AI scientists, right up to the present day.

TURING'S INTELLIGENCE ERROR AND NARROW AI

The problem-solving view of intelligence helps explain the production of invariably narrow applications of AI throughout its history. Game playing, for instance, has been a source of constant inspiration for the development of advanced AI techniques, but games are simplifications of life that reward simplified views of intelligence. A chess program plays chess, but does rather poorly driving a car. IBM's Watson system plays *Jeopardy!*, but not chess or Go, and massive programming or "porting" efforts are required to use the Watson platform to perform other data mining and natural language processing functions, as with recent (and largely unsuccessful) forays into health care.

Treating intelligence as problem-solving thus gives us narrow applications. Turing no doubt knew this, and speculated in his 1950 paper that perhaps machines could be made to learn, thus overcoming the constraints that are a natural consequence of designing a computer system narrowly to solve a problem. If machines could learn to become general, we would witness a smooth transition from specific applications to general thinking beings. We would have AI.

What we now know, however, argues strongly against the learning approach suggested early on by Turing. To accomplish their goals, what are now called machine learning systems must each learn something specific. Researchers call this giving the machine a "bias." (This doesn't carry the negative connotation it does in the broader social world; it doesn't mean that the machine is pigheaded or difficult to argue with, or has an agenda in the usual sense of the word.) A bias in machine learning means that the system is designed and tuned to learn something. But this is, of course, just the problem of producing narrow problem-solving applications. (This is why, for example, the deep learning systems used by Facebook to recognize human faces haven't also learned to calculate your taxes.)

Even worse, researchers have realized that giving a machine learning system a bias to learn a particular application or task means it will do more poorly on other tasks. There is an inverse correlation between a machine's success in learning some one thing, and its success in learning some other thing. Even seemingly similar tasks are inversely related in terms of performance. A computer system that learns to play championship-level Go won't also learn to play championship-level chess. The Go system has been specifically designed, with a particular bias toward learning the rules of Go. Its learning curve, as they call it, thus follows the known scoring of that particular game. Its learning curve regarding some other game, say *Jeopardy!* or chess, is useless—in fact, nonexistent.

Machine learning bias is typically understood as a source of learning error, a technical problem. (It has also taken on the secondary meaning, hewing to ordinary language usage, of producing results that are unintentionally and unacceptably weighted by, say, race or gender.) Machine learning bias can introduce error simply because the system doesn't "look" for certain solutions in the first place. But bias is actually necessary in machine learning—it's part of learning itself.

A well-known theorem called the "no free lunch" theorem proves exactly what we anecdotally witness when designing and building learning systems. The theorem states that any bias-free learning system will perform no better than chance when applied to arbitrary problems. This is a fancy way of stating that designers of systems must give the system a bias deliberately, so it learns what's intended. As the theorem states, a truly bias-free system is useless. There are complicated techniques, like "pre-training" on data using unsupervised methods that expose the features of the data to be learned. All of this is part and parcel of successful machine learning. What's left out of the discussion, however, is that tuning a system to learn what's intended by imparting to it a desired bias generally means causing it to become narrow, in the sense that it won't then generalize to other domains. Part of what it means to build and deploy a successful machine learning system is that the system is not bias-free and general but focused on a particular learning problem. Viewed this way, narrowness is to some extent baked in to such approaches. Success and narrowness are two sides of the same coin.

This fact alone casts serious doubt on any expectation of a smooth progression from today's AI to tomorrow's human-level AI. People who assume that extensions of modern machine learning methods like deep learning will somehow "train up," or learn to be intelligent like humans, do not understand the fundamental limitations that are already known. Admitting the necessity of supplying a bias to learning systems is tantamount to Turing's observing that insights about mathematics must be supplied by human minds from outside formal methods, since machine learning bias is determined, prior to learning, by human designers.[10]

TURING'S LEGACY

To sum up the argument, the problem-solving view of intelligence necessarily produces narrow applications, and is therefore inadequate for the broader goals of AI. We inherited this view of intelligence from Alan Turing. (Why, for instance, do we even use the term artificial intelligence, rather than, perhaps, speaking of "human-task-simulation"?)[11] Turing's great genius was to clear away theoretical obstacles and objections to the possibility of engineering an autonomous machine, but in so doing he narrowed the scope and definition of intelligence itself. It is no wonder, then, that AI began producing narrow problem-solving applications, and it is still doing so to this day.

Turing, again, disliked viewing thinking or intelligence as something social or situational. Yet despite his proclivities to see human intelligence as an individual mechanical process—ushering in untold media references to the "mechanical brain" as early computers appeared in the 1940s—it is obvious that talk of intelligence always involves, as it must necessarily involve, situating it in a broader context. General (non-narrow) intelligence of the sort we all display daily is not an algorithm running in our heads, but calls on the entire cultural, historical, and social context within which we think and act in the world. AI would hardly have moved forward if developers had embraced such a large and complicated understanding of intelligence—that is true enough. At the same time, as a result of Turing's simplification, we've ended up with narrow applications, and we have no reason to expect general ones without a radical reconceptualization of what we mean by AI.

Turing anticipated some of these difficulties in his 1950 paper by suggesting that machines might be made to learn. What we now know, however (contra recent excitement about machine learning), is that learning itself is a kind of problem-solving, made possible only by introducing a bias into the learner that simultaneously makes possible the learning of a particular application, while reducing performance on other applications. Learning systems are actually just narrow problem-solving systems, too. Given that there is no known theoretical bridge from such narrow systems to general intelligence of the sort displayed by humans, AI has fallen into a trap. Early errors in understanding intelligence have led, by degrees and inexorably, to a theoretical impasse at the heart of AI.

Consider again Turing's original distinction between intuition and ingenuity. The question of AI for him was whether intuition—that which is supplied by the designer of a system—could in fact be "pulled into" the formal part of the system (the ingenuity machine), thus making a system capable of escaping the curse of narrowness by using intuition to choose its own problems—to grow smarter and to learn. So far, no one has done this with any computer. No one even has the slightest clue how this would work. We do know that designers use intuition outside AI systems to tell such systems what specific problems to solve (or to learn to solve). The question of systems using intuition autonomously goes straight to the core of what I will call the Problem of Inference, to which we will turn in Part Two.

There will also be much more to say about the "narrowness trap" of AI in Part Two. First, however, there is more ground to cover in this part. We will turn next to superintelligence, another intelligence error, and a natural extension of the first.

Chapter 3

• • •

THE SUPERINTELLIGENCE ERROR

Jack Good, Turing's fellow code-breaker, also became fascinated with the idea of smart machines. Turing no doubt primed his colleague's imagination at Bletchley and afterward, and Good added a sci-fi-like twist to Turing's ideas about the possibility of human-level intelligence in computers. Good's idea was simple: if a machine can reach human-level intelligence, it can also surpass mere human thinking.

Good thought it obvious that a feedback loop of sorts would enable smart machines to examine and improve themselves, creating even-smarter machines, resulting in a runaway "intelligence explosion." The explosion of intelligence follows because each machine makes a copy of itself that is still smarter—the result is an exponential curve of intelligence in machines that quickly surpass even human geniuses. Good called it *ultraintelligence*: "Let an ultraintelligent machine be defined as a machine that can far surpass all the intellectual activities of any man however clever. Since the design of machines is one of these intellectual activities, an ultraintelligent machine could design even better machines; there would then unquestionably be an 'intelligence explosion,' and the intelligence of man would be left far behind. Thus the first ultraintelligent machine is the last invention that man need ever make, provided that the machine is docile enough to tell us how to keep it under control."[1]

Oxford philosopher Nick Bostrom would return to Good's theme decades later, with his 2014 best seller *Superintelligence: Paths, Dangers, Strategies,* making the same case that the achievement of AI would as a consequence usher in greater-than-human intelligence in an escalating process of self-modification. In ominous language, Bostrom echoes Good's futurism about the arrival of superintelligent machines:

> Before the prospect of an intelligence explosion, we humans are like small children playing with a bomb. Such is the mismatch between the power of our plaything and the immaturity of our conduct. Superintelligence is a challenge for which we are not ready now and will not be ready for a long time. We have little idea when the detonation will occur, though if we hold the device to our ear we can hear a faint ticking sound. For a child with an undetonated bomb in its hands, a sensible thing to do would be to put it down gently, quickly back out of the room, and contact the nearest adult. Yet what we have here is not one child but many, each with access to an independent trigger mechanism. The chances that we will all find the sense to put down the dangerous stuff seem almost negligible. Some little idiot is bound to press the ignite button just to see what happens.[2]

To Bostrom, superintelligence is not speculative or murky at all, but rather like the arrival of nuclear weapons—a fait accompli, and one which has profound and perhaps dire consequences for mankind. The message here is clear: don't dispute whether superintelligence is coming. Get ready for it.

What are we to say about this? The Good-Bostrom argument—the possibility of a superintelligent machine—seems plausible on its face. But unsurprisingly, the mechanism by which "super" intelligence results from a baseline intelligence is never specified. Good and Bostrom seem to take the possibility of superintelligence as obviously plausible and therefore not requiring further explanation. But it does; we do need to understand the "how."

If we suppose a simple enhancement like superior hardware, the proposal is too trivial and silly to entertain further. Even a stalwart believer in inexorable progress like Ray Kurzweil isn't likely to reduce intelligence that far—we don't think adding RAM to a MacBook makes it (really and truly) more intelligent. The device is now faster, and can load bigger applications, and so on. But if intelligence means anything interesting, it must be more complicated than loading applications faster. This harder part of intelligence is left unsaid.

Or suppose we borrow language from the biological world (as AI so often does), and then confidently declare that computational capability doesn't devolve, it evolves. Looking deeper, we see that this argument is plagued once again by an inadequate and naive view of intelligence. The problem—a glaring omission—is that we have no evidence in the biological world of anything intelligent ever designing a more intelligent version of itself. Humans are intelligent, but in the span of human history we have never constructed more intelligent versions of ourselves.

A precondition for building a smarter brain is to first understand how the ones we have are cognitive, in the sense that we can imagine scenarios, entertain thoughts and their connections, find solutions, and discover new problems. Things occur to us; we reason through our observations and what we already know; answers pop into our heads. All of this buzz of biological magic remains opaque, its "processing" still vastly uncharted. And yet, we have been contemplating and investigating our thinking processes and brain functions for millennia.

Why should a generally intelligent machine suddenly have insight into its own global cognitive capacities, when we clearly do not? And even if it did, how could the machine use this knowledge to make itself smarter?

This is not a matter of self-improvement. We can, for instance, make ourselves more intelligent by reading books or going to school; educating ourselves makes possible further intellectual development, and so on. All of this is uncontroversial. And none of it is the point. One major problem with assumptions about increases in intelligence in AI circles is the problem of circularity: it takes (seemingly general) intelligence to increase general intelligence. A closer look reveals no linear progression, but only mystery.

VON NEUMANN AND SELF-REPRODUCING MACHINES

Good introduced the idea of self-improving AIs leading to ultraintelligence in the mid-1960s, but nearly two decades earlier John Von Neumann had considered the idea and rejected it. In a 1948 talk at the Institute for Advanced Studies at Princeton, Von Neumann explained that, while human reproduction often improves on prior "designs," it's clear that machines tasked with designing new and better machines face a fundamental stumbling block, since any design for a new machine must be specified in the parent machine. The parent machine would then necessarily be more complex, not less, than its creation: "An organization which synthesizes something is necessarily more complicated, of a higher order, than the organization it synthesizes," he said.[3]

In other words, Von Neumann pointed to a fundamental difference between organic life as we know it, and the machines we build. Jack Good's prediction of ultraintelligence was a bit of science fiction.

Von Neumann theorized that a self-reproducing machine would need, at minimum, eight parts, including a "stimulus organ," a "fusing organ" to connect parts together, a "cutting organ" to sever connections, and a "muscle" for motion. He then sketched plausible mechanisms for cognitive improvements including a randomizing element, akin to biological mutation, to allow for the necessary modifications. But Von Neumann thought that, rather than advance the machine's thinking, such random mutations were more likely to "devolve" desired functions and capacities. The most probable outcome was nonfunction, the equivalent of a lethal change: "So, while this system is exceedingly primitive, it has the trait of an inheritable mutation, even to the point that a mutation made at random is probably lethal, but may be non-lethal and inheritable."

For the machines to get something better, essentially smarter, from their designs they would need a creative element added to their stimulus and fusing organs. Unlike biological evolution, the idea wasn't to wait around millions of years, but to require of parent systems the necessary Promethean spark in themselves, leading more or less directly to better designs. This was fiction, thought Von Neumann. As he explained to his colleagues at Princeton, no science or engineering theories could make sense of it. Von Neumann, no Luddite, was exploding the "intelligence explosion."

One obvious flaw in predictions of an intelligence explosion leading to superintelligence is that we already have human-level intelligence—we are human. By Good's logic, we should then be capable of designing something better than human. This is just a restatement of the goals of the field of AI, so we are getting trapped in a circle. The humans who are AI researchers already know it's a mystery how to design smarter artifacts, just as Von Neumann explained. Transferring this mystery from our own intelligence to an envisioned machine's doesn't help. To unpack this more, consider a genius AI researcher we'll call Alice.

INTELLIGENCE EXPLOSIONS, THE VERY IDEA

Let's suppose Alice is an AI scientist who has a dull neighbor, Bob. Bob has common sense, can read the newspaper, and can carry on a conversation (although perhaps it's boring), so he's worlds ahead of the best AI systems coming out of Google's DeepMind.

Alice works for an amazing new startup (soon to be acquired by Google), and wants to build an AI that is as smart as Bob. She's sketched out two systems, in the spirit of Daniel Kahneman's well-known System 1 and System 2.[4] They are intuition pumps or metaphors that give a rough blueprint for the types of problems needing to be solved to get to artificial general intelligence. In Alice's context, we'll call these System X, for competence on well-defined tasks like game play (as in chess or Go) and System Y, for general intelligence. The latter system includes Bob's competence at reading and conversation, but also the murkier area of novel ideas and insights.

Bob is terrible at chess, and in fact his X system is pathetic compared not only to a system like AlphaGo but also to many other humans. His short-term memory is worse than most people's; he scores poorly on IQ tests; and he struggles with crossword puzzles. As for his Y system, his general intelligence shows a conspicuous lack of interest in or ability at novel or insightful thinking. Bob is not the kind of neighbor that gets many invitations to dinner parties.

Alice's strategy is first to design a Bob-Machine that matches Bob's intelligence. She reasons that if she succeeds in creating a Bob-Machine, that machine can design a smarter version of itself, leading eventually to an intelligence explosion. Now, again, keep in mind that designing a Bob-Machine is no easy task, because Bob has a System Y—which means he has solved the problem of commonsense reasoning and has general cognitive abilities. He can pass a Turing test, for one. And he can read children's stories and the sports section and summarize them. Bob therefore blows away Google's best natural language understanding systems, like Ray Kurzweil's Talk to Books semantic search tool. This is why Alice is excited about her Bob-Machine project; it would be a huge advance in AI.

The question is: how to get there? Alice's first approach is to maximize the Bob-Machine's System X capabilities. She gives it a computer memory and access to the web via Google. Unfortunately, this version of the Bob-Machine quickly proves Stuart Russell's point that supercomputers without real intelligence just get to wrong answers more quickly.[5] The Bob-Machine remembers the wrong things and fails to ask the right questions. All the improvements on the System X side just make the machine more competent at recalling and coughing up crackpot theories and making pronouncements about the world with more facts, all misused and poorly understood from a System Y perspective. Sure, the Bob-Machine plays flawless chess, but its chess competence makes it less interesting to Alice, who realizes that the machine she has created has no hope of designing a "more intelligent" version of itself.

In an *aha!* moment, though, Alice realizes that Bob himself couldn't design a smarter version of himself. So how could the Bob-Machine? The problem, she thinks, is that System X optimization does not supply resources to System Y of the necessary kind. The Bob-Machine (like Bob himself) has to see its own intelligence as something of a certain quantity, assess how it is limited and to what extent, and then actively redesign itself so as to become smarter in the important and relevant ways. But this is precisely the way in which the Bob-Machine (like Bob) is unintelligent! The Bob-Machine *can't* do this, because it lacks these System Y capabilities for insight, discovery, and innovation. Alice must go back to the drawing board.

Alice then decides that the Bob-Machine is just too stupid to be part of a bootstrapping process to superintelligence. (In a moment of sheer panic, it occurs to her that this logic jeopardizes the entire enterprise of getting to superintelligence, but she manages to suppress this concern quickly.) Alice decides, in deference to AI's founder and to the eager-beaver marketing department of her company, Ultra++, that she'll instead focus on designing a machine as intelligent as Alan Turing, called the Turing-Machine.

Now, assuming that Turing was smarter than Alice (though who is to say?), she can't just design a Turing-Machine directly, and anyway she already ran into a brick wall puzzling out how to design a Bob-Machine. She decides to begin with a machine that's as smart as Hugh Alexander—Turing's colleague at Bletchley Park and a one-time chess champion of Cambridge. Hugh Alexander was smart—really smart. He played championship-level chess, and although he did not summon the kinds of insights about breaking the Enigma code that Turing did, he did make valuable contributions and earn the respect of the other Bletchley code breakers—no small feat. The Hugh-Machine should be smart enough to figure out how to wire a Turing-Machine, and a machine at Turing's level must certainly be smart enough to make itself even smarter!

Alice easily succeeds at improving Hugh-Machine's chess competence (despite his being a champion player already), by simply downloading some StockFish chess code off her smartphone. Similarly, she gives the Hugh-Machine perfect arithmetical abilities with a calculator, and supercomputer memory, as well as access to all the information retrievable by Google. System X is optimal, and the Hugh-Machine can do something that Hugh Alexander, for all his seeming intelligence, could not: it can play superhuman chess, and superhumanly add numbers, and excel at many other System X things. The problem is, so too could Bob-Machine. In fact, Alice realizes that Bob-Machine and Hugh-Machine are provably equivalent. In fact, she is forced to admit (over several glasses of red wine) that abandoning Bob-Machine was pointless and self-defeating.

Another glass of wine and a cigarette later, Alice turns off her phone to silence annoying text messages from Ultra++ colleagues about her impending breakthrough. The truth is, she muses, that it's not just Bob that can't design a smarter version of himself, it's Alice, too. In a moment of clarity, she grasps that the further we move from System X toward System Y, toward insight and innovation, the more opaque the design will become. Turing, for instance, could judge his intelligence at chess—losing to both Hugh Alexander and Jack Good.

But Turing could not assess his own System Y capabilities. In a very real sense, his intellect was a black box, and he could not at any rate evaluate his own competence at original thinking (whatever that means), not only because it is not a blueprint, as it were, but because it sits in time, in a lifetime, and might still produce new as yet unpredictable ideas. Turing's System Y intelligence is not just unpredictable, in other words, it's inexplicable—perhaps not to some being smarter than Turing (again: whatever that means) but certainly to Turing himself. Same for dullard Bob. How then would Alice ever ignite an intelligence explosion?

In fact, the very idea of an intelligence explosion has built into it a false premise, easily exposed once someone ambitious and insightful like Alice first takes it seriously. By hypothesis, a Bob-Machine is as intelligent as Bob. Well, here's an idea. Go ask Bob to design a slightly smarter version of himself. You will find this is not something Bob can do. The essential quality of mind that makes AI exciting also forecloses the linear assumption of an intelligence explosion. "Once we get to human-levels of intelligence, the system can design a smarter-than-human version of itself," so the hope goes. But, we already *have* "human-level" intelligence—we're human. Can *we* do this? What are the intelligence explosion promoters really talking about?

This is another way of saying that the powers of the human mind outstrip our ability to mechanize it in the sense necessary for "scaling up," from AlphaGo to a Bob-Machine to a Turing-Machine, and beyond. The intelligence explosion idea itself is not a particularly good System Y candidate for progress on AI toward general intelligence.

THE EVOLUTIONARY TECHNOLOGISTS

Many AI enthusiasts who hold to an inevitability thesis (superintelligent machines are coming, no matter what we do) hold to this because it plays on evolutionary themes, and thus conveniently absolves individual scientists from the responsibility of needing to make scientific breakthroughs or develop revolutionary ideas. Artificial intelligence just evolves, like we did. We can call the futurists and AI believers in this camp evolutionary technologists, or ETs.

The ET view is popular among new age technologists like *Wired* cofounder Kevin Kelly, who argues in his 2010 book *What Technology Wants* that AI won't come about as the work of a "mad scientist," but simply as an evolutionary process on the planet, much like natural evolution.[6] According to this view, the world is becoming "intelligenized" (Kelly's word), and more and more complex and intelligent forms of technology are emerging without explicit human design.[7] Such thinkers also might envision the World Wide Web as a giant, growing brain. Humans, in this view, become a link in a cosmic historical chain that reaches into the future to true AI, where we get left behind or assimilated.

Organic life evolves extremely slowly, but ETs view technological progress as accelerating. As Kurzweil famously argues, technology is getting more complicated on an accelerating curve, according to a law he thinks is discernible in history, the Law of Accelerating Returns. Thus, human-level intelligence and then superintelligence will emerge on the planet in drastically short timeframes, as compared to organic evolution. In decades or even years, we will be confronted with them.

This is a simple, tidy story of humanity. We are transitioning to something else, which will be smarter and better.

Notice that the story is not testable; we just have to wait around and see. If the predicted year of true AI's coming is false, too, another one can be forecast, a few decades into the future. AI in this sense is unfalsifiable and thus—according to the accepted rules of the scientific method—unscientific.

Note that I'm not saying that true AI is impossible. As Stuart Russell and other AI researchers like to point out, twentieth-century scientists such as Ernest Rutherford thought that building an atomic bomb was impossible, but Leo Szilard figured out how nuclear chain reactions work—a mere twenty-four hours after Rutherford pronounced the idea dead.[8] It's a good reminder not to bet against science. But note that nuclear chain reactions grew from scientific theories that were testable. Theories about mind power evolving out of technology aren't testable.

The claims of Good and Bostrom, presented as scientific inevitability, are more like imagination pumps: just think if this could be! And there's no doubt, it would be amazing. Perhaps dangerous. But imagining a what-if scenario stops far short of serious discussion about what's up ahead.

For starters, a general superintelligence capability must be connected to the broader world in such a way that it can observe and "guess" more productively than we do. And if intelligence is also social and situational, as it seems it must be, then an immense amount of contextual knowledge is required to engineer something more intelligent. Good's problem is not narrow and mechanical, but rather pulls into its orbit the whole of culture and society. Where is the barest, even remotely plausible blueprint for this?

Good's proposal, in other words, is based once again on an inadequate and simplified view of intelligence. It presupposes the original intelligence error, and adds to it yet another reductive sleight of hand: that an individual mechanical intelligence can design and construct a greater one. That a machine would be situated at such an Archimedean point of creation seems implausible, to put it mildly. The idea of superintelligence is in reality a multiplication of errors, and it represents in barest form the extension of the fantasy about the rise of AI.

To dig deeper into all of this, we should push further into this fantasy. It's called the *Singularity,* and we turn to it next.

Chapter 4

• • •

THE SINGULARITY, THEN AND NOW

In the 1950s, the mathematician Stanislaw Ulam recalled an old conversation with John Von Neumann, in which Von Neumann discussed the possibility of a technological turning point for humanity: "the ever accelerating progress of technology . . . gives the appearance of approaching some essential singularity in the history of the race beyond which human affairs, as we know them, could not continue."[1]

Von Neumann likely made this comment as digital computers were arriving on the technological scene. But digital computers were the latest innovation in a long and seemingly unbroken sequence of technologies.[2] By the 1940s, it had become clear that the scientific and industrial revolutions of the past three hundred years had set in motion forces of immense, symbiotic power: the fruits of new science seeded the development of new technology, which in turn made possible more scientific discovery. For example, science gave us the telescope, which in turn improved astronomy.

Inextricably tied to changes in science and technology was social change—rapid, chaotic at times, and seemingly irreversible. City populations exploded (with considerable doses of squalor and injustice), and entirely new forms of social and economic organization emerged, seemingly overnight. Steam engines revolutionized transportation, as did, later, internal combustion engines. Trains, trolleys, and steamboats opened up trade, and migration into cities created entirely new workforces. With Thomas Edison's invention of the electric lightbulb, people could work at night; insomniacs in rural areas could now read *Das Kapital* or *On the Origin of the Species* after the sun had abandoned them. Productivity soared. Wealth and health increased. So did blood and violence. A sequence of geopolitical events led to "the Great War," World War I, which introduced chemical warfare on a mass scale. And a couple of decades later, in Von Neumann's world, the ultimate existential threat—the nuclear bomb—had become a reality.

The bomb was an inflection point in history, making clear the dystopian possibilities inherent in unbridled technological innovation. Shannon and Turing were using computers to play chess; scientists like Von Neumann were using computers to develop weapons to vaporize Japanese cities. Electronic computers were big and slow, but were still orders of magnitude faster than human computers at tasks like calculating numeric progressions, which Von Neumann and others used to determine likely nuclear blast radii given different quantities of fissile material.

It was in this miasma of possibilities and fears that Von Neumann posed the question of a "singularity." Famously polymathic and brilliant, Von Neumann was almost universally respected by his scientific colleagues, including Alan Turing, and it is unsurprising that his suggestion affected Ulam, who remembered it decades later.

Ulam the mathematician no doubt understood Von Neumann's metaphor. A singularity is a mathematical term, indicating a point which becomes undefined—a value which, say, explodes to infinity. Von Neumann asked Ulam whether the trajectory of technological progress would in effect approach "infinity," where no methods or thoughts, strategies or actions could be applied. Prediction would become impossible. Progress would no longer be a known variable (if it ever was).

Von Neumann, in other words, suggested to Ulam an eschatology, a possible end of times. A couple decades later, Good thought he'd found the mechanism: the digital computer.

UCLA computer scientist and Hugo award-winner Vernor Vinge introduced "Singularity" into computation, and specifically into artificial intelligence, in 1986, in his science-fiction book *Marooned in Realtime*.[3] In a later technical paper for NASA, Vinge channeled Good: "Within thirty years, we will have the technological means to create superhuman intelligence. Shortly after, the human era will be ended. . . . I think it's fair to call this event a singularity. It is a point where our models must be discarded and a new reality rules. As we move closer and closer to this point, it will loom vaster and vaster over human affairs till the notion becomes a commonplace. Yet when it finally happens it may still be a great surprise and a greater unknown."[4]

Vinge, the computer scientist, had professional company. By the late 1980s MIT computer scientist, futurist, and entrepreneur Raymond Kurzweil had become AI's "bulldog," spreading the Singularity idea into pop culture science in a series of publications, beginning with *The Age of Intelligent Machines* in 1990, then in 1998 his follow-up *The Age of Spiritual Machines.* His 2005 best seller was even more confident: *The Singularity is Near.*

Kurzweil argued that technological innovations, when plotted on a historical graph, are exponential. Innovation accelerates, viewed historically, as a function of time. In other words, the time between major technological innovations keeps shrinking. For example, paper appeared in the second century, and the printing press took another twelve hundred years—Gutenberg's press appeared in Europe in 1440. But computation appeared in the 1940s (the 1930s if one counts its mathematical treatment), and the internet—quite a major innovation—showed up less than thirty years later. And AI? By Kurzweil's logic, human-level intelligence in the computer is mere decades away—maybe less. Exponential growth curves surprise us.

Kurzweil branded this "The Law of Accelerated Returns" (LOAR), and used it as a premise in an argument whose conclusion was that fully-human AI would arrive by 2029, and then through a process of bootstrapping more intelligent machines, superintelligence by 2045.[5]

Superintelligence signaled the point of no return, where the path of progress disappears into the unknown, the Singularity. This is the crossover point, where machines and not people take over as the most intelligent beings on the planet.

Kurzweil famously views this process as entirely "scientific," citing LOAR (though LOAR is not a law at all) and in no small measure his own verve and credentials as a computer expert and inventor (Kurzweil helped develop text-to-speech technologies, leading to modern systems like Siri).

Turing. Good. Vinge. The ideas about radical change made possible by advances in computation were already in the air. Kurzweil ostensibly provided the roadmap. Like so many others obsessed with the question of AI, his prose brimmed with all the zeal of the converted:

> We are entering a new era. I call it "the Singularity." It's a merger between human intelligence and machine intelligence that is going to create something bigger than itself. It's the cutting edge of evolution on our planet. One can make a strong case that it's actually the cutting edge of the evolution of intelligence in general, because there's no indication that it's occurred anywhere else. To me that is what human civilization is all about. It is part of our destiny and part of the destiny of evolution to continue to progress ever faster, and to grow the power of intelligence exponentially. To contemplate stopping that—to think human beings are fine the way they are—is a misplaced fond remembrance of what human beings used to be. What human beings are is a species that has undergone a cultural and

> technological evolution, and it's the nature of evolution that it accelerates, and that its powers grow exponentially, and that's what we're talking about. The next stage of this will be to amplify our own intellectual powers with the results of our technology.[6]

In truth, however, by the time Kurzweil so enthusiastically jumped on board—indeed, decades before the 1980s—work on scientific AI itself had extinguished hopes of any inexorable march to superintelligence. Research and development on actual AI had proved, in a word, difficult.

By the 1970s, AI gadfly and MIT philosopher Hubert Dreyfus had published an influential dismissal of the field as an egregious example of what Hungarian philosopher Imre Lakatos referred to as a "degenerating research program."[7] Dreyfus was bombastic, but he had a point, as actual computer scientists knew all too well. The field was suffering one setback after another, with well-funded efforts and grandiose claims about intelligent machines consistently (and often quite dramatically) falling short of expectations. Research labs at MIT, Stanford, and elsewhere were encountering seemingly endless quandaries, difficulties, confusions, and outright failures. The problem, for example, of Fully Automated High-Quality Machine Translation was thought in the 1950s to be solvable, given sufficient research effort and dollars. By the 1960s, government investment in translation dried up, on the heels of one failure after another. Hopes of building robots with common sense (say, abilities to understand and speak English) also evaporated—or at least were drastically dimmed by a stream of early disappointments. Conversational systems that were intended to take realistic Turing tests succeeded in fooling human questioners only with tricks and bluffing—not real understanding of language—a problem that still bedevils natural language efforts in AI today. AI seemed inevitable in press releases and talk of the future, but not when the focus was on actual work in research labs. Programming a truly intelligent machine turned out to be hard. Really hard.

As the idea of the Singularity gained traction in pop culture, AI scientists kept wandering into seemingly endless engineering problems. And so the sky was not falling. The Singularity was not near. Vinge's popular fiction remained just that.

A close inspection of AI reveals an embarrassing gap between actual progress by computer scientists working on AI and the futuristic visions they and others like to describe. Turing, in 1950, had proposed a question to be tested: Can machines be smart like people? Good, Vinge, Kurzweil, and others have answered the question with a resounding *yes* without taking seriously the actual nature of the problems encountered by real work in the field.

This gap is instructive.

In particular, the failure of AI to make substantive progress on difficult aspects of natural language understanding suggests that the differences between minds and machines are more subtle and complicated than Turing imagined. Our use of language is central to our intelligence. And if the history of AI is any guide, it represents a profound difficulty for AI.

We turn to it next.

Chapter 5

• • •

NATURAL LANGUAGE UNDERSTANDING

Artificial intelligence as an official discipline began, auspiciously, in 1956 at the now-famous Dartmouth Conference. Luminaries in attendance included Shannon of Bell Labs (information theory), Marvin Minsky of Harvard (mathematics), noted Carnegie Mellon economist Herbert Simon, John McCarthy, Harvard psychologist George Miller (known for his work on human memory), and John Nash (the Nobel laureate mathematician famously portrayed in the 2001 movie *A Beautiful Mind*).

McCarthy, then at Dartmouth but soon to take a position in the new field of computer science at Stanford, coined the term artificial intelligence at the conference, giving a name, officially, to the modern project of engineering intelligent life. Back in 1816, a precocious young Mary Shelley had begun work on her masterpiece *Frankenstein*. A hundred and forty years later, the scientists assembled at Dartmouth were contemplating the assembly of a new "modern Prometheus," which would soon burst out into the public view.

The field was hyped from the get-go. The conference proceedings themselves said it all:

> We propose that a 2 month, 10 man study of artificial intelligence be carried out during the summer of 1956 at Dartmouth College in Hanover, New Hampshire. The study is to proceed

> on the basis of the conjecture that every aspect of learning or any other feature of intelligence can in principle be so precisely described that a machine can be made to simulate it. An attempt will be made to find how to make machines use language, form abstractions and concepts, solve kinds of problems now reserved for humans, and improve themselves. We think that a significant advance can be made in one or more of these problems if a carefully selected group of scientists work on it together for a summer.[1]

The agenda at Dartmouth was simple: investigate the nature of cognitive (thinking) capabilities, design programs to reproduce those capabilities, and implement and test their performance on the new electronic computers. As the Dartmouth participants made clear, in the summer of 1956, with ten researchers armed with knowledge in their respective scientific fields, they expected a "significant advance" toward engineering human intelligence on a machine.

Working at RAND, Herbert Simon and Allan Newell designed AI programs in the late 1950s that seemed to make good on the bullish Dartmouth Conference promises. The AI program Logic Theorist, and later the General Problem Solver, used a simple heuristic search to prove theorems of traditional logic and to solve logic-based puzzles in clear, computational steps. The programs worked, and AI seemed destined to unlock the secrets of human intelligence quickly, just as the Dartmouth organizers had declared.

Initial successes by Simon and Newell quickly emboldened researchers to set more ambitious goals. Turing had already set the endgame agenda a decade earlier with his version of the Imitation Game, the Turing test. Dartmouth scientists, too, thought that programming a machine to understand English or any other natural language would constitute a declaration of victory for AI. Researchers had long thought that natural language understanding was "AI-Complete," lingo borrowed from mathematics to mean that at the point computers tamed natural language they would have achieved general intelligence, and thus be capable of thinking and acting like humans. By the 1960s, therefore, the target of AI was the task of machine translation—the fully automatic rendering of texts from one language, such as Russian, into another, such as English. AI was "all in."

NATURAL LANGUAGE UNDERSTANDING

As AI turned to natural language understanding, its practitioners radiated a confidence in imminent success that continued the tradition begun at Dartmouth. Herbert Simon, who would go on to win the prestigious A. M. Turing Award and then the Nobel Memorial Prize in Economics, announced in 1957 that "there are now in the world machines that think, that learn and that create." In 1965, he prognosticated that, by 1985, "machines will be capable of doing any work Man can do." Marvin Minsky, too, declared in 1967 that "within a generation, the problem of creating 'artificial intelligence' will be substantially solved."[2]

But machine translation was a different ballgame, as researchers soon discovered. Having begun with a simplistic assumption, that language could be understood by analyzing words in large texts (called *corpora*) using statistical techniques, they were quickly proven wrong. Computers made automatic translation possible, but the results were far from high quality. Even programs working in specific domains such as biomedical literature were not fail-proof, and the failures were often embarrassingly incorrect and mindless.

Machine translation researchers, in response, expanded their approach by exploring methods for "parsing" sentences, or finding syntactic structure in them, using new and powerful "transformational" grammars developed by a young MIT linguist who was soon to be world-famous—Noam Chomsky. But extracting correct parses from natural language texts itself proved vastly more difficult and complex than anyone had imagined. Problems surfaced that, in retrospect, should have been obvious. These included word-sense ambiguity (by which a word like *bank* carries different possible meanings); non-local contextual dependence (by which a word's meaning depends on other words in a discourse or text that are not in its immediate vicinity); and other issues involving reference (anaphor), metaphor, and semantics (meaning). As philosopher and cognitive scientist Jerry Fodor put it, AI had walked into a game of three-dimensional chess thinking it was tic-tac-toe.[3] The National Resource Council (NRC) was pouring millions into machine-translation work at a number of American universities by the mid-1960s, but as for actual successes in engineering systems to understand, or even to simulate understanding of, natural language texts, they were, to put it mildly, in short supply.

MIT researcher Yehoshua Bar-Hillel, once a fiercely enthusiastic supporter of fully automated translation, was first to sound an alarm bell. He did more than that, in fact, in a series of official reports for the NRC and in now-famous footnotes spelling out the depth of problems the field faced.[4] The effects of his reports on the research community were seismic. He had pinpointed the exact obstacle machine translation was foundering on, and it was irritatingly "philosophical": the dearth of so-called common sense or "world knowledge"—knowledge about the actual world. Consider a simple sentence: *The box is in the pen.* Bar-Hillel explained that it confounded automated systems, no matter how sophisticated, because they lacked a simple, actual knowledge of the world. Knowledge about the relative sizes of pens and boxes enables humans to disambiguate such sentences almost instantly. We quickly recognize that the pen in question is not likely a writing instrument, but rather an enclosure for young children or animals. And it becomes all the more obvious with some additional context, as in Bar-Hillel's example: *Little John was looking for his toy box. Finally he found it. The box was in the pen. John was very happy.* But automated systems lacking such knowledge face a mysterious, seemingly impossible task.

World knowledge, as Bar-Hillel pointed out, couldn't really be supplied to computers—at least not in any straightforward, engineering manner—because the "number of facts we human beings know is, in a certain very pregnant sense, infinite."[5] He had discovered, unwittingly, that humans know much more than anyone had imagined—the opposite of a quick and simple solution for AI. And it was the apparently mundane, commonsense, quotidian facts about the everyday world that tripped up the most sophisticated automated systems. Any seemingly ordinary fact could become relevant in the course of a translation, yet making the requisite, open-ended quantities of "knowledge" accessible to computational systems in real-time or near real-time, and imbuing them with cognitive abilities to select relevant facts against this open-ended (possibly infinite) background, seemed hopeless. As Bar-Hillel concluded in his notorious 1966 report to the NRC, the notion that computers could be programmed with the world knowledge of humans was "utterly chimerical, and hardly deserves any further discussion."[6]

Machine translation was stuck, in other words, with results that were a far cry from fully automatic, high-quality translations (and that remain so today, although the quality has improved). Thus the pattern continued. AI had oversold itself, and in the wake of the failure of the translation research to live up to promises, the NRC pulled its funding after investing over twenty million dollars into research and development, an enormous sum at the time. In the wake of the debacle, AI researchers lost their jobs, careers were destroyed, and AI as a discipline found itself back at the drawing board.[7]

Attempts to tame or solve the "commonsense knowledge problem" dominated efforts in AI research in the 1970s and 1980s. By the early 1990s, however, AI still had no fresh approaches or answers to its core scientific—and philosophical—problem. Japan had invested millions in its high-profile Fifth Generation project aimed at achieving success in robotics, and Japan too had failed, rather spectacularly. By the mid-1990s, AI found itself again in a "winter"—no confidence in the promises of AI researchers, no results to prove naysayers wrong, and no funding. Then came the web.

THE WORLD WIDE WEB

The emergence of the World Wide Web spurred resurgence in AI for a simple reason: data. Suddenly, the availability of massive datasets, and in particular of text corpora (web pages) from the combined efforts of millions of new web users, breathed life into old, "shallow" statistical and pattern-recognition approaches. Suddenly, what used to be shallow became adequate and worked. Supervised learning algorithms such as artificial neural networks (neural nets, for short), decision trees, and Bayesian classifiers had existed in university labs for decades. But without large datasets, they hadn't yet shown much promise on interesting problems like face recognition, or text classification, or spam or fraud detection. Such methods now seemed filled with endless promise—and for real-world, moneymaking applications that would bring a fresh wave of attention and funding to AI.

And so Big Data was born (the term came a bit later). By the turn of the century, the so-called shallow, ground up, empirical, or data-driven approaches typified by learning algorithms such as neural nets and graphical models had opened up vast opportunities in both AI research and AI business applications. New methods were developed—incorporating hidden Markov models, maximum entropy models, conditional random fields, and large-margin classifiers such as support vector machines—and rapidly came to dominate pure and applied research in AI. Seemingly overnight, a whole science of statistical and numerical analysis appeared, based on optimizing learning methods operating on Big Data. Universities launched projects in natural language understanding and natural language processing. They found ways to, for example, extract names and other patterns from web pages (a capability called entity recognition); to disambiguate polysemous (multi-sense) words such as *bank*; to perform web-specific tasks like ranking and retrieving web pages (the famous example being Google's PageRank, which Larry Page and Sergey Brin developed as Stanford graduate students in the 1990s); to classify news stories and other web pages by topic; to filter spam for email; and to serve up spontaneous product recommendations on commerce sites like Amazon. The list goes on and on.

The shift away from linguistics and rule-based approaches to data-driven or "empirical" methods seemed to liberate AI from those early, cloudy days of work on machine translation, when seemingly endless problems with capturing meaning and context plagued engineering efforts. In fact, machine translation itself was later cracked by a group of IBM researchers using a statistical (that is, not grammar-based) approach that was essentially an ingenious application of Claude Shannon's early work on information theory. Called the "noisy channel" approach, it viewed sentences from a source language (say, French) and a target language (say, English) as an information exchange in which bad translations constituted a form of noise—making it the system's task to reduce the noise in the translation channel between source and target sentences. The idea worked, and machines began producing usable translations using the vastly simpler—though data intensive—statistical approach pioneered by IBM Research Labs.

SUCCESS . . . OR NOT

The success of contemporary systems like Google Translate on the once puzzling problem of machine translation is often touted as evidence that AI will succeed, given enough time and the right ideas. The truth is more sobering.

While it turns out that some problems in natural language understanding can be addressed with statistical or machine learning approaches, the original concerns of Bar-Hillel and others regarding semantics (meaning) and pragmatics (context) have proven to be well-founded. Machine translation, which had seemed like a difficult natural language problem, could be adequately accomplished with simple statistical analysis, given large corpora (datasets) in different languages. (And note that machine translation is still not very high quality—it is more like "good enough to be useful.") This is not evidence of impressive growth in machines' natural language understanding intelligence, but only evidence that machine translation is a much simpler problem than it was initially perceived to be.

Again, deep problems with understanding language using computers have persisted. A simple way to see the point here is to turn back to the Turing test, and reconsider it in light of the history of AI and the many, mostly fruitless attempts to solve the problems it presents or even to make substantive progress. Futurists like Nick Bostrom, as well as the larger scientific AI community, likely wish that the public would just forget about the test.

It's not—as is sometimes heard—that the test is flawed or unhelpful. Quite simply, it's too hard.

THE TURING TEST

Given a view from thirty thousand feet, it did seem that computers were getting smarter as AI progressed from its genesis in Turing's early work and the kickoff conference at Dartmouth. Certainly computers were gaining more powerful processors and storage capabilities, and these led inevitably to progress on well-defined problems. Computer chess-play did not yet exceed skilled humans, but kept improving. Programs like DENDRAL, which analyzed the structure of chemicals, and MYCIN, which provided sometimes quite good medical diagnoses, made clear that AI methods were relevant to a variety of problems normally requiring high human intelligence. Machine translation, as we've seen, was an initial failure, but yielded to different approaches made possible by the availability of large datasets (a precursor to many Big Data and data science successes in the 2000s). All sorts of natural language processing tasks, like generating parses of natural language sentences, and tagging parts of speech or entities (persons, organizations, places, and the like), were chipped away at by AI systems with increasing power and sophistication.[8]

Yet Turing's original goal for AI, passing the Turing test, remained elusive. Understanding conversation was not a well-defined task for AI—it was an entire capability. While there was some hope of success with early systems like Terry Winograd's SHRDLU system, which stacked and manipulated blocks with a robotic arm by parsing English-language commands like *Place Block A on top of B, and put them both on top of C*, researchers drew blanks on open-ended conversational systems. Progress amounted to trickery, to clever ploys by designers of systems to fool human interlocutors by feigning ignorance or adopting evasive responses and strategies when questions put the programs out of their depth. Perhaps the most famous trick system, Joseph Weizenbaum's ELIZA program, could create the illusion of human dialogue by mimicking a Rogerian psychotherapist's prompts to patients. Here is a sample, with ELIZA's responses in italics:

> Men are all alike.
>
> *In what way*
>
> They're always bugging us about something or other.
>
> *Can you think of a specific example*
>
> Well, my boyfriend made me come here.

Your boyfriend made you come here

He says I'm depressed much of the time.

I am sorry to hear you are depressed

It's true. I am unhappy.

Do you think coming here will help you not to be unhappy

I need some help, that much seems certain.[9]

But such programs were obvious frauds, judged by the standards of true AI.

So was the program that on multiple occasions briefly fooled some judges at the Loebner Prize contest, an annual staging of the Turing test. Media reports eagerly declared that the test had finally been passed, and there was brief but intense fanfare about the progress of AI on its most famous challenge. The chatbot—named Eugene Goostman—was presented as a thirteen-year-old Ukrainian boy, a persona designed to make judges expect (and excuse) grammatical errors and holes in general knowledge. The bot also responded to some questions not with straightforward answers, but with misdirection and deflection in the form of random comments and humorous asides. Did the bot trick some judges some of the time? Yes. Did it actually pass the Turing test in any meaningful way? No.[10]

Though machine translation has in recent years been tamed (somewhat) by large volumes of texts translated into different languages on the web, the Turing test remains a perpetual frustration for AI. The ghost of Bar-Hillel still haunts us.

Chapter 6

• • •

AI AS TECHNOLOGICAL KITSCH

In 1980, the Czech-born writer Milan Kundera wrote his masterpiece, *The Unbearable Lightness of Being.* The novel is a love story, set against the backdrop of the Soviet invasion of then Czechoslovakia by the Soviet Union in 1968. Kundera wrote about the writers and artists who committed suicide after relentless, mendacious harassment by the Soviet secret police who had inserted themselves into the social, intellectual, and cultural fabric of Prague. Dead and discredited, the Prague intellectuals then received further (though posthumous) disgrace: disgusting encomiums at their funerals, where Soviet party members and officials would attest to the deceased's lifelong devotion to the State. Soviet propaganda drove them to death; the same propaganda then portrayed their lives as nobly sacrificed to advance ideas that they had in fact spoken out against publicly and privately. What they hated, they were described as loving.

The Soviet propaganda was ruthless, but it was not wrathful and stupid. It had a particular purpose. That purpose was to purge the country of deeper and more profound (and contrary) expressions of the meaning of a country, a people, and a life. The Soviets were purging Prague, and all of Czechoslovakia, of its shared history, its traditions, and its sense of what was valuable and worth fighting for. Once the free-thinkers were silenced, the Soviets would then, like painting a wall after first sandblasting it, be free to impose their worldview without serious or organized opposition. Kundera's story is a trenchant and often tragic account of the value of human life, and how particular beliefs and ideologies can attempt—but never quite manage—to obfuscate and gloss over all that is meaningful to an individual and to a society. Kundera called the Soviet culture foisted upon the defeated Czech people *kitsch*.

TECHNOLOGICAL KITSCH

Kitsch is a German word that, while it usually refers today to cheesy or tacky artwork and decor, originally meant exaggerated sentimentality and melodrama in any realm. The intelligence errors at the heart of the AI worldview—the beliefs, that is, not the science—have given rise to a modern and particularly pernicious form of kitsch. Dreams of superintelligent computers are not Soviet propaganda, and no one is coercing us to believe in the rise of the machines. But they share the basic idea of replacing complex and difficult discussions about individuals and societies with technological stories that, like Soviet culture, rewrite older ideas with dangerously one-dimensional abstractions.

Kitsch is a word whose meaning and use have changed over time. The original German definition in some ways differs from the meaning I intend to explore here, but two essential ingredients of the original meaning should make my claim clear enough. First, kitsch involves a simplification of complicated ideas. There must be a simple story to tell. Second, it offers easy solutions that sweep away, with emotion, the questions and confusions people have about the problems of life rather than addressing those questions with serious, probing discussion. Thus, a perfect example of kitsch is the dreamy idea that one day an awe-inspiring android with superintelligence will remake human society and its older traditions and ideas, and we'll enter a new era, thankfully free of old arguments about God, mind, freedom, the good life, and the like. Beautiful machines (or machines with beautiful intelligence) like "Ava" in the 2015 sci-fi film *Ex Machina*, portrayed by Alicia Vikander, will remove the hard facts of human existence. This is kitsch, technological-style. Like Soviet propaganda, it might horrify or mollify, but it gives us a new story that writes over and makes unnecessary what was true before, and the old reality disappears.

Alan Turing, for all his contributions to science and engineering, made possible the genesis and viral growth of technological kitsch by first equating intelligence with problem-solving. Jack Good later compounded Turing's intelligence error with his much-discussed notion of ultraintelligence, proposing that the arrival of intelligent machines necessarily implied the arrival of superintelligent machines. Once the popular imagination accepted the idea of superintelligent machines, the rewriting of human purpose, meaning, and history could be told within the parameters of computation and technology.

But ultraintelligent machines are fanciful, and pretending otherwise encourages the unwanted creep of technological kitsch, usually in one of two ways that are equally superficial. At one extreme we hear a tale of apocalyptic or fearsome AI, a sort of campfire horror story. At the other extreme we encounter utopian or dreamy AI, which is equally superficial and unmerited. If we take either form of AI's kitsch seriously, we end up in a world defined only by technology.

This is a theme I will be returning to, because it exposes the core problem with futuristic AI. As Nathan, the genius computer scientist in *Ex Machina* puts it, "One day the AIs are going to look back on us the same way we look at fossil skeletons on the plains of Africa. An upright ape living in dust with crude language and tools, all set for extinction." In truth, it's unclear that any computer will ever look back at all. The popular sentiment requires a deep dive into the meaning of existence, life, consciousness, and intelligence, and the differences between ourselves and computation and its many technologies. Kitsch prevents us from grappling with human nature and other serious philosophical endeavors. This simply shouldn't be the case, as Kundera knew all too well.

Kitsch has its roots, typically, in a larger system of thought. For the communists, it was Marxism. With the inevitability myth, it's technoscience. We inherited the technoscientific worldview most directly from the work of August Comte.

OUR TECHNOSCIENTIFIC CONDITION

Probably the first thinker to fully develop and explain technoscience as a worldview was Auguste Comte.[1] Comte, the nineteenth-century philosopher widely credited with founding sociology as a scientific field of study, developed and expounded the theory of positivism. This is the view that only observable, scientific phenomena exist—religion and philosophy are imaginary. Comte made explicit, first, his idea that the human mind progresses toward truth, as does society as a whole, through stages that begin with religious and philosophical thinking and then advance to scientific thinking. And, second, he explained that technoscience would eventually create a heaven on earth by enabling the nature of all things (science) to be understood, and then using this knowledge to develop technologies that make our lives vastly longer, better, and more meaningful.

Comte's account of the transformative power of technoscience extended, eventually, into his belief that religion and in particular the Church could be replaced by a "religion of humanity" which would be thoroughly secular, believing in no God and grounded firmly in the sciences and in material reality. At the time of Comte's writing in the nineteenth century, there was sufficient evidence of the power of human thinking both to discover scientific laws and to innovate powerful and useful technologies that technoscience took root at the center of the modern mind.

From the beginning, however, there were misgivings about Comte's theory. Nietzsche, for instance, lamented that the idea of a person became constricted and limited with such a view. Technoscience might help us live longer, but it could not make us wiser. The idea of a hero or a person of extraordinary and earned gifts and virtues didn't fit Comte's vision, which had essentially replaced traditional discussion of personhood with discussion of the progress of science, and especially technology.[2]

Comte's materialism, too, suggested to other thinkers a diminution rather than expansion of human possibility. To the east, in Russia, the writer Dostoevsky protested contemptuously about the growing "scourge" of total belief in materialism and scientism—the view that scientific knowledge is the only real knowledge—in prose that reflected other thinkers' skepticism and even fear of the rapid dominance of technoscientific thinking. As he put it in his *Notes from Underground*, "One's own free and unfettered volition, one's own caprice, however wild, one's own fancy, inflamed sometimes to the point of madness—that is the one best and greatest good, which is never taken into consideration because it cannot fit into any classification, and the omission of which sends all systems and theories to the devil."[3]

Dostoevsky, Nietzsche, and others were pointing to the ideal of a full person, but Comte was talking about the ideal of something external to us—about technoscience and its advance. The problem was that, as Comte well knew, the vision of a technoscientific future was also a deep and significant statement about the nature of personhood. In effect, Comte argued that traditional conceptions of personhood—as unique because created by God, or as a seeker of wisdom (not only technological knowledge) as the Greek philosophers held—were by virtue of scientific and technological success now irrelevant. His technoscience philosophy was commentary on the essence and possibilities of human nature. This was radical, and iconoclastic thinkers not taken in by the juggernaut of technoscience were right to challenge the notions that Comte (and others) propounded.[4]

THE TRIUMPH OF *HOMO FABER*

Technoscience triumphed in the twentieth century but skeptical responses to it continued, as well. Hannah Arendt, the philosopher made famous by her phrase "the banality of evil," in reference to the

Nazi Nuremberg trials, argued that Comte's technoscience—which, by the middle of the twentieth century, certainly had not lost any steam as a philosophical idea—amounted to no less than a redefinition of human nature itself.[5] Arendt pointed to the classical understanding of humans as *Homo sapiens*—literally, wise man—and to the historical focus on wisdom and knowledge rather than technical skill, and argued that to embrace technoscience as a worldview was to redefine ourselves as *Homo faber*—man the builder.

Homo faber, in Greek terms, is a person who believes that *techne*—knowledge of craft or making things, the root of technology—defines who we are. The faberian understanding of human nature fits perfectly not only with Comte's nineteenth-century idea of a utopian technoscience but with the twentieth-century obsession with building more and more powerful technologies, culminating in the grand project of, in effect, building ourselves—artificial intelligence. This project would not make sense if the traditional notions of the meaning of humanity had remained intact.

Arendt argued that the seismic change from wisdom and knowledge to technology and building represented a limiting and potentially dangerous understanding of ourselves, which would guarantee not only that technological development would continue unbridled, but that increasingly we would view technological successes as meaningful statements about ourselves. We were, in other words, reducing our own worth in order to increase, beyond wise or reasonable measure, our estimation of the marvels that could be built with the tools of technoscience.

Von Neumann's initially cryptic comments about approaching a "singularity" as technological advances accelerate become more clear in light of his contemporary Arendt's position. Though Von Neumann, a scientist and mathematician, did not (as far as we know) further explain his remarks, they perfectly reflect Arendt's insistence on the deep significance of technoscience for ourselves and our future—for what philosophers of technology call "the human condition." It would perhaps seem perverse to Comte that technology could accelerate past our control, but nowhere in his writing can one discover an inkling of the point that Arendt (and others) would make, that in championing technoscience as a human answer to human problems, we are also engaged in the project of redefining our understanding of ourselves. The turn toward techne rather than, say, *episteme* (knowledge of natural phenomena) or *sapientiae* (wisdom relating to human values and society) makes it difficult to carve out a meaningful idea of human uniqueness. (Even bees, after all, are builders, in their case of hives).

Putting techne at the center also makes it possible to view a person as something that can be built, since it implies there is nothing more to a person than a superior capacity to construct ever more advanced technologies. Once embarked on this route, it is a short journey to artificial intelligence. And here is the obvious tie-in with the intelligence errors first made by Turing and then extended by Jack Good and others up to the present day: the ultimate triumph of *Homo faber* as a species is to build itself. This is, of course, precisely the professed goal of AI. Exploring whether the project can succeed or not will necessarily pull us into the deep waters of understanding the nature of ourselves.

FILLING IN THE PUZZLE

Technoscience began with the Scientific Revolution, and by a few hundred years later much of modern scientific theory was in place. With rare exceptions—an obvious one being the development of quantum theory and relativity in the twentieth century—scientific knowledge advanced as major physical theories were put in place. Scientific knowledge was like a puzzle, with pieces of theory forming a picture of the world and the universe. Newton's physics, Maxwell's electrodynamics, theories of work and thermodynamics from Carnot and others—all of this scientific knowledge fit together to form a unified picture of the world. More theories and details were filled in by

Darwin's theory of evolution in the 1850s and by geographical and archeological discoveries. (Of course, as theories were debated and tested, some proved wrong or were revised). The range of possibility for scientific theory, then, was strangely shrinking—as when one is working a jigsaw puzzle, and with each piece fitted into place the remaining choices are further limited.

By contrast, technological innovation exploded. As Ray Kurzweil has noted, technological innovation accelerates. One invention does not limit what can follow, but rather makes possible more and more inventions. Technology seems, in other words, to evolve. We do not put it in place like theory. Instead, we accrete technological developments one on top of the other, seemingly endlessly. The acceleration of the evolution of technology means simply that the time between major technological innovations keeps shrinking, historically, so that the gap between, say, the invention of the printing press and the arrival of the computer is very large compared to, say, the gap between the computer and the internet. The merger of science and technology is thus complicated, and the very word *technoscience* suggests that, as things progress, science will settle and technology will continue to evolve—and to evolve, as Kurzweil puts it, exponentially.

And so the term *technoscience* itself demonstrates the complexity and unpredictability of our world. Not all areas of human endeavor follow the same pattern of growth; one area can't be laid along another, as with a template. Whether human intelligence and machine intelligence are more alike than not—or more unalike—remains to be seen. The question of AI should be an invitation not to ignore philosophical conundrums but to struggle with them. And technoscience, taken as a statement about ourselves, is in the end a terrible simplification. It represents (among other things) the introduction of kitsch into the stream of complicated and difficult issues in life.

Chapter 7

• • •

SIMPLIFICATIONS AND MYSTERIES

Shortly before Turing published his 1950 "Computing Machinery and Intelligence," the behavioral psychologist B. F. Skinner published a science fiction novel, *Walden Two.*[1] In it, Skinner has his characters argue that free will is an illusion, and that a person's behavior is controlled from the outside, by his or her environment. If someone (a scientist, say) changes the environment, then the behavior of the person in that environment will change.

In a trivial sense, this is true. If a despot denies food, security, and opportunities for employment from people, the people will grow unhappy. We can predict such changes. Skinner meant, however, that a person is entirely determined by inputs—in his terms, by stimuli.

In fact, Skinner's idea of a person as a "black box" is the basic idea Turing had in mind, too. With a black box, we treat the output of the system as some function of its input—the *how* of the internals in the system itself is left undescribed. Skinner argued in *Walden Two* that a perfect world—a utopia—could be constructed by treating people like black boxes, by feeding them certain physical input (stimuli) to achieve a certain output (response). Meanwhile, Turing speculated that the human being was operationally equivalent to some complicated machine, and to prove this, he suggested building a machine, feeding it input, and examining its output.

Much of importance was left out of this way of thinking, unfortunately, and it seems obvious today that we have inherited its mistakes. While Skinner's theory of operant conditioning, or "behaviorism," as it came to be called, was widely disputed later in the twentieth century, the interdisciplinary "cognitive revolution" that replaced it treated intelligence as merely internal computations. This idea, underpinned by a philosophy called the "computational theory of mind," which claims that the human mind is an information processing system, still underwrites theoretical confidence in the eventual triumph of AI.

Here, it's best to be clear: equating a mind with a computer is not scientific, it's philosophical.

THE FOLLY OF PREDICTION

As Stuart Russell points out, in the quest for artificial intelligence, we shouldn't bet against "human ingenuity."[2] But in a similar vein we shouldn't make hopeful (or dire) predictions without a sound scientific basis.

Experts and even (or especially) scientists love to make predictions, but most of them are wrong. Dan Gardner's excellent book *Future Babble* documents the success rate of predictions in realms from history and geopolitics to the sciences.[3] He found that theorists—experts with big visions of the future based on a particular theory they endorse—tend to make worse predictions than pragmatic people, who see the world as complicated and lacking a clear fit with any single theory.

Gardner referred to the expert class and the pragmatic thinkers as hedgehogs and foxes (borrowing from Philip Tetlock, the psychologist, who himself borrowed the terminology from Isaiah Berlin). Just as a hedgehog burrows into the ground, hedgehog experts burrow into an idea. Inevitably they come to believe that the idea captures the essence of everything, and that belief fuels their inevitable proselytizing. Marx was a tireless hedgehog.

Foxes see complexity and incalculability in the affairs of the world, and either avoid bold predictions or make the safer (and perhaps smarter) prediction that things won't change the way we think. For the fox, the business of predicting is almost foolhardy, because we really can't know what will emerge from the complicated dynamics of geopolitics, domestic politics (say: who will win an election?), science, and technology. As the nineteenth-century novelist Leo Tolstoy warned, wars unfold for reasons that we can't fit into battle plans.

Some AI scientists are notoriously foxy about AI predictions. Take Yoshua Bengio, a professor of computer science at the University of Montreal, Canada, and one of the pioneers of deep learning: "You won't be getting that from me," he says, in response to the question of when we can expect human-level AI: "there's no point. It's useless to guess a date because we have no clue. All I can say is that it's not going to happen in the next few years."[4]

Ray Kurzweil gives a more hedgehog answer: human-level AI will arrive in 2029. He invokes his "law" of accelerating returns to make his prediction seem scientific, and he sees continuing evidence that he's right in all the supposed progress to date.[5]

Philosophers sometimes have the virtue of thinking clearly about problems precisely because they are unencumbered by any particular zeal that might attach itself to practitioners in a field (who wish still to philosophize). Alasdair MacIntyre, for example, in his now classic *After Virtue,* pointed to four sources of fundamental unpredictability in the world. In particular, his discussion of "radical conceptual innovation" is directly germane to questions about when human-level AI will arrive. He recalls the argument against the possibility of predicting invention made by twentieth-century philosopher of science Karl Popper:

> Some time in the Old Stone Age you and I are discussing the future and I predict that within the next ten years someone will invent the wheel. "Wheel?" you ask. "What is that?" I then describe the wheel to you, finding words, doubtless with difficulty, for the very first time to say what a rim, spokes, a hub and perhaps an axle will be. Then I pause, aghast. "But no one can be *going* to invent the wheel, for I have just invented it." In other words, the invention of the wheel cannot be predicted. For a necessary part of predicting an invention is to say what a wheel is; and to say what a wheel is just *is* to invent it. It is easy to see how this example can be generalized. Any invention, any discovery, which consists essentially in the elaboration of a radically new concept cannot be predicted, for a necessary part of the prediction is the present elaboration of the very concept whose discovery or invention was to take place only in the future. The notion of the prediction of radical conceptual innovation is itself conceptually incoherent.[6]

In other words, to suggest that we are on a "path" to artificial general intelligence whose arrival can be predicted presupposes that there is no conceptual innovation standing in the way—a view that even AI scientists convinced of the coming of artificial general intelligence and who are willing to offer predictions, like Ray Kurzweil, would not assent to. We all know, at least, that for any putative artificial general intelligence system to arrive at an as yet unknown facility for understanding natural language, there must be an invention or discovery of a commonsense, generalizing component. This certainly counts as an example of a "radical conceptual innovation," because we have no idea what this is yet, or what it would even look like.

The idea that we can predict the arrival of AI typically sneaks in a premise, to varying degrees acknowledged, that successes on narrow

AI systems like playing games will scale up to general intelligence, and so the predictive line from artificial intelligence to artificial general intelligence can be drawn with some confidence. This is a bad assumption, both for encouraging progress in the field toward artificial general intelligence, and for the logic of the argument for prediction.

Predictions about scientific discoveries are perhaps best understood as indulgences of mythology; indeed, only in the realm of the mythical can certainty about the arrival of artificial general intelligence abide, untrammeled by Popper's or MacIntyre's or anyone else's doubts.

Mythology about AI is not all bad. It keeps alive archetypal longings for creating life and intelligence, and can open windows into understanding ourselves. But when myth masquerades as science and certainty, it confuses the public, and frustrates non-mythological researchers who know that major theoretical obstacles remain unsolved. "No one has a clue," as Bengio puts it. This is impossibly and depressingly pessimistic for mythologists, even if supported by all the evidence, and true.

Obstacles are not always insurmountable, however, and even when they are—when we are forced to recognize certain boundaries—we are then freed to find a different way to reach our goal, or given the impetus to formulate new goals altogether. The history of science is chock-full of examples of the discovery of impasses leading to further progress. Werner Heisenberg discovered his uncertainty principle by working out the consequences of the new physics of quanta. The principle states that it is impossible to isolate the position and the momentum of a subatomic particle simultaneously. This places fundamental limits on our ability to predict the individual movements of particles at the subatomic realm (because "seeing" the position of a particle requires impinging it with a photon, which also has the effect of knocking it off course). The uncertainty principle is nothing if not a limitation, yet it has proven fruitful and valuable in understanding quantum mechanics. We could not, for instance, ever hope to build a quantum computer if we first didn't understand the nature of uncertainty.

There are many other examples. Perpetual motion was an obsession of the nineteenth and eighteenth centuries, pulling into its orbit many of the best and brightest minds. Advances in theories of work and thermodynamics retired the dream—but in the process allowed for huge progress in understanding energy and motion. Admitting complexity—and complications—gets us further than does easy oversimplification.

A STRANGE (BUT RELEVANT) ARGUMENT FROM MICHAEL POLANYI

One possibility in the AI debate is that we have general intelligence, but we can't actually write down what it is—program it, that is—because in important respects it's a black box to ourselves. This brings us to Michael Polanyi.

Once influential but now little-known, chemist and philosopher Michael Polanyi argued in the mid-twentieth century that intelligence is only partly captured by the symbols we write down—the uses of language that he called "articulations." Polanyi was anticipating many of the headaches AI systems have caused for AI designers; in fact, in his later works he explicitly denied that machines could capture all of human intelligence, for reasons stemming from the incompleteness of articulations.

Polanyi argued that articulations necessarily leave out "tacit" components of intelligence—aspects of thinking that can't be precisely described by writing down symbols.[7] (A neural network that we construct is a symbol system, too.) This explains why, for instance, certain skills and crafts, like cooking, can't be mastered by simply reading recipes. We make things, but this doesn't mean we can program everything we make (think of writing a program for writing a novel on the order of, say, James Joyce's *Ulysses*. The program would be meaningless. Instead we would write the novel directly—if we were James Joyce).

Polanyi wrote at an unfortunate time to suggest contrary views about AI, as the field had kicked off in the 1950s with much fanfare. His defense of tacit knowledge was picked up later in the previously discussed attack on AI by Hubert Dreyfus; perhaps because of his sometimes too tendentious tone, Dreyfus's remarks became a lightning rod for counterarguments, and at least initially, did not win over mainstream AI thinkers. (Unfortunately, he also declared that an AI system could never beat a grand champion at chess.)[8]

But the possibility that not all of what we know can be written down is an enduring problem for AI, because it implies that AI programmers are attempting to square a circle. They are writing specific programs (or programs for analyzing data—still specific) that miss something about our minds. Polanyi's ideas suggest that minds and machines have fundamental differences, and also that equating minds with machines leads to a simplification of our ideas about the mind. If the mind—or at least general intelligence—must be treated as something that can be coded or written down, then we must simplify "mind" itself to make sense of so much discussion today.

A RETURN OF THE FOXES

In the early 2000s, everyone in AI was a fox. The field was experiencing one of its perennial winters, and most all the mythologists were in hiding. Ray Kurzweil still promoted his confident vision, and classical AI theorists like Doug Lenat kept pursuing their favorite theories, chasing the Rosetta stone of AI. But seemingly endless boom-and-bust cycles had worn down much of the field, to the point where many felt uncomfortable even using the label of AI for our research. It became a bad marketing term. (That seems strange today, but it was true.) Talk turned quickly to the arcana of specific algorithms, like "support vector machines" and "maximum entropy," both approaches to machine learning. Classical AI scientists dismissed these as "shallow" or "empirical," because statistical approaches using data didn't use knowledge and couldn't handle reasoning or planning very well (if at all). But with the web providing the much-needed data, the approaches started showing promise.

The deep learning "revolution" began around 2006, with early work by Geoff Hinton, Yann LeCun, and Yoshua Bengio. By 2010, Google, Microsoft, and other Big Tech companies were using neural networks for major consumer applications such as voice recognition, and by 2012, Android smartphones featured neural network technology. From about this time up through 2020 (as I write this), deep learning has been the hammer causing all the problems of AI to look like a nail—problems that can be approached "from the ground up," like playing games and recognizing voice and image data, now account for most of the research and commercial dollars in AI.

As deep learning took off, AI (and talk of AI) did too. Hedgehogs returned, and predictably, the media fanned the flames of fresh futurism. But something strange is happening in AI lately. I noticed it in more skeptical talk in 2018, and in 2019 it's unmistakable. The foxes are returning.

Many mythologists (with a few notable exceptions) are also nonexperts, like Elon Musk, or the late astrophysicist Stephen Hawking, or even Bill Gates. Still, they helped create much of the media ballyhoo about AI—mostly, deep learning ballyhoo—which peaked a few years ago (circa 2015, give or take a year). Now, though, it's increasingly common to hear talk of limitations again—for instance, from Gary Marcus, a cognitive scientist and founder of robotics company Robust.AI, who coauthored with computer scientist Ernest Davis the 2019 *Rebooting AI: Building Artificial Intelligence We Can*

Trust.[9] Marcus and Davis make a compelling argument that the field is yet again overhyped, and that deep learning has its limits; some fundamental advance will be required to achieve generally intelligent AI. In 2017, AI scientist Hector Levesque (a colleague of Davis, about whom more will be said later) penned a helpful polemic about modern AI which he titled *Common Sense, the Turing Test, and the Quest for Real AI*.[10] Back when I published "Questioning the Hype about Artificial Intelligence" in *The Atlantic* in 2015, reactions were largely dismissive.[11] Today there are more critics, and among them are many recognized leaders in AI who are questioning the hype.

It's still rare to hear thoughtful arguments that true AI is impossible, for the same reason people shy away from offering predictions about it—because the future of AI is an unknown. But culturally and psychologically, the field seems to have entered a phase of dialing down, cautioning newbies and the expectant public that general intelligence is a long road. This trend is of utmost importance, because the myth is an emotional lighthouse by which we navigate the AI topic. It's expansionary, inviting all comers: concepts like consciousness, emotions like aggression or love, instincts like sex, and other ingredients of minds and living beings. But the new "science" talk is, more or less, a narrative about possible extensions of narrow AI to more and more generality, where big-picture ideas like consciousness are out of scope. Too clever by half, perhaps—the myth is why everyone cares. Otherwise, it's just more powerful forms of technology everywhere, a trend we can already see is double-sided.

SIMPLIFYING SUPERINTELLIGENCE

The intelligence errors that helped forge our simplified computational world are now back in a modern guise, as well. Stuart Russell, who coauthored the definitive textbook introduction to AI with Google's

Peter Norvig, argues in his 2019 *Human Compatible: Artificial Intelligence and the Problem of Control* that intelligence means nothing more than achieving objectives—providing a definition that includes not only humans and dolphins but also ants, *E. coli* bacteria, and computers.[12] Also, he wants the Turing test to be retired because it's now irrelevant. (Apparently, having an ordinary conversation is not a worthy objective.) "The Turing test is not useful for AI," he writes, "because it's an informal and highly contingent definition: it depends on the enormously complicated and largely unknown characteristic of the human mind, which derives from both biology and culture. There is no way to 'unpack' the definition and work back from it to create machines that will provably pass the test. Instead, AI has focused on rational behavior [and thus] a machine is intelligent to the extent that what it does is likely to achieve what it wants, given what it has perceived."[13]

It's hard to argue with Russell's definition of intelligence, which covers everything from Einstein "achieving" his "objective" when he reimagined physics as relativity, to a daisy turning its face toward the sun. But Russell's dismissal of the Turing test seems overly legalistic and narrow, because the spirit of the test is simply that machines that understand and use natural languages must be intelligent. Practically speaking, we shouldn't expect much from Siri or other voice-activated personal assistants if they never figure out what we're saying, so dismissal of the test seems unwise. (If some next-generation Siri ever advances to the point where it engages in unrestricted and ordinary conversation with its human owner, then the Turing test will return as the great "dream of AI," finally realized. Alas.)

Russell, a recognized AI expert and professor of computer science at the University of California, Berkeley, also rids himself of the problem of consciousness: "In the area of consciousness, we really do know nothing, so I'm going to say nothing." He then assures us that "No one in AI is working on making machines conscious, nor would anyone know where to start, and no behavior has consciousness as a prerequisite." But he says something—quite a lot—about consciousness anyway:

> Suppose I give you a program and ask, "Does this present a threat to humanity?" You analyze the code and indeed, when run, the code will form and carry out a plan whose result will be the destruction of the human race, just as a chess program will form and carry out a plan whose result will be the defeat of any human who faces it. Now suppose I tell you that the code, when run, also creates a form of machine consciousness. Will that change your prediction? Not at all. It makes absolutely no difference. Your prediction about its behavior is exactly the same, because the prediction is based on the code. All those Hollywood plots about machines mysteriously becoming conscious and hating humans are really missing the point: it's competence, not consciousness, that matters.[14]

But maybe it's Russell who is missing the point, because mythology about machines "coming alive" is really the lifeblood of dreams about future AI. If we were to inform folks arriving at the theater for a screening of *Ex Machina* that the real dream of AI was to make mindless, "no lights on inside" supercomputers to help us (and our enemies) achieve objectives, they might feel a bit underwhelmed. Russell seems to suggest that an algorithmic system, suitably juiced up with as yet unknown modules for general intelligence, will spell ultimate success for AI. True hedgehogs understand what cautionary foxes do not, that AI straddles science and myth, and its enduring fascination in the popular mind means it's a psychological and cultural touchstone.

Ray Kurzweil has argued all along that, whatever consciousness is, machines will have it in spades—richer and "better" than our own.

His 1999 paean to the myth was appropriately titled *The Age of Spiritual Machines* (and he really means "spiritual," as in superintelligent computers having conscious, spiritual experiences). Kurzweil wisely insists that Turing's test is a proper benchmark for AI: "In order to pass the test, you have to be intelligent." He even dislikes the term "AGI," increasingly used to specify artificial general intelligence, because as he puts it (correctly), "the goal of AI has always been to achieve greater and greater intelligence and ultimately to reach human levels of intelligence.[15]

Sexual desire might even be a proper subject for AI as a litmus test for intelligence—especially if it's a basic element in striving and yearning, and achieving various objectives, as it seems to be in life. *Ex Machina* is practically Shakespearean, combining sexual tension, consciousness, exploitation, and liberation—and all in a Turing test (of sorts). Novelist-turned-director Alex Garland tackles the Singularity, giving us the story of a superintelligent android masterminding her escape from enslavement by a mad scientist (another pregnant theme)—her reclusive inventor, Nathan (played by Oscar Isaac). Ostensibly, Ava's objective is to pass a Turing test by interacting with Nathan's invited guest, Caleb Smith (Domhnall Gleeson), as a thoroughly convincing "human"—even though he has been informed and can see that Ava is an android. It's the ultimate test, says Nathan.

But Ava has her own ideas (achieving our objectives be damned), and plots an escape into the wild world, outside the confines of Nathan's research facility. When Ava finally escapes, two humans are (or will be) dead. As she steps out, she sees color, glorious color—proof to the viewer that she's really "alive" and conscious.[16] We have seen her understand and use the English language so effectively that she has reduced the two men to hopeless confusion and defeat. Here, then, is a full-throated depiction of the myth, presenting its core futuristic idea of a coming crossover point when machines overtake humans full-stop. Ava is smarter, more cunning, and more spiritual and alive than her human counterparts.

Garland's vision is pure mythology—and also a great human story, capturing archetypal themes (liberation, good and evil, and sexuality) through the lens of future technology. There is an irony here, because *Ex Machina* succeeds by tapping into deep human emotions, as do other masterpieces like *2001: A Space Odyssey.*

Also ironic is the choice some have made recently to distance the field of AI from its joyful (or fearful) myth in favor of more "scientific" discussion—in other words, to dismiss emotionally fraught concepts like the Singularity, consciousness, and intelligence while still benefiting from their remaining in the public eye. Russell, for instance, clearly wants to separate serious work on artificial general intelligence from pop cultural portrayals of it in movies like *Ex Machina.* He regards consciousness as a silly philosophical worry (no one knows anyway), Turing tests as antiquated ideas too vulnerable to tricks, and any worries about machines getting aggressive (or plucky, or otherwise emotional) as fundamentally misguided. Superintelligent computers will simply pursue their objectives. The problem is—and it amounts to an existential risk, even sans *Terminator* imagery—that their objectives might not be our own.

Russell admits that this is our problem with AI already. Specifically, he calls out those content-selection algorithms on the web whose objective is to maximize ad revenue by bombarding everyone with sticky and relevant ads. Superintelligent AI may be too good at pursuing our objectives. The apt metaphor is the story of King Midas, who gained the power to turn anything to gold, but found it too easy to turn everything to gold, including his own daughter (not the objective); similarly, the superintelligence we charge with an objective might find a way of achieving it that ends up eliminating us, perhaps even by using the carbon atoms in human beings themselves as further resources, as means to its end.

The idea is a perennial worry among superintelligence worriers. Nick Bostrom considers a scenario where superintelligence is assigned the seemingly mundane task of maximizing the production of paper clips (its human-given objective), and by degrees converts everything in the universe into a paper clip factory, including all the usable elements in our own bodies. Eliezer Yudkowsky, former head of Berkeley's Machine Intelligence Research Institute, once quipped, "The AI does not hate you, nor does it love you, but you are made out of atoms which it can use for something else."[17]

The idea that the coming superintelligence will somehow be laser-focused and uber-competent at achieving an objective yet have zero common sense seems to cut against the grain of superintelligence itself—which is, after all, supposed to be human intelligence plus more. AI scientists like Gary Marcus, who understand intelligence as (minimally) having common sense (and perhaps have some themselves) point out that a superintelligent computer optimizing manufacture of a human product for sale, like paper clips, might also hit upon the idea that it shouldn't destroy the humans that buy them. Again, there's a curious simplification of superintelligence implicit in Russell's and others' worries about it acting as a superpowered automaton with blind computational adherence to an objective given to it by its programmers. It's an odd position to stake out. Russell himself admits that common sense and language is a major and unachieved milestone for AI. Why is it absent in his picture of superintelligence? Future computers possessing common sense would obviate such worries—unless they were aggressive and diabolic after all, which Russell is at pains to dismiss as silly myth.

At any rate, paper clip apocalypse scenarios *do* bother scientifically-minded researchers like Russell, who suggest we forestall such possibilities by building into future superintelligent computers certain principles—first, to ensure that they "attach no intrinsic value" to their own well-being, having the sole objective of maximizing *our* preferences. The problem, as Russell reminds us, is that we are often clueless about our own preferences. At the very least, we are apt to misstate what we want, in the King Midas sense.

Thus, as another principle in addition to altruism toward humanity, AI must also be imbued with humility, to thwart any errors it might make in pursuing altruism toward us (like converting the paper clip factory CEO into a paper clip, thinking it's really, really squeezing every ounce of productivity, using all possible means). Altruistically humble machines help guard against the danger that cigar-smoking tech executives (who probably don't smoke cigars anymore) could give them venal motives, and also against the possibility of machines being too smart in the wrong way, doing the equivalent of turning everything into gold. Confusingly, "altruistically humble" machines also sound a lot like the *Ex Machina* take on AI—as "alive" after all, with real (not just paper clip maximizing) intelligence and ethical sensibilities. One might be forgiven for drawing the conclusion that talk of AI is doomed perpetually to straddle science and myth.

Russell has a third principle he thinks necessary to thwart existential crisis with the coming superintelligence: AI should be developed in such a way that it learns to predict human preferences. Machines should watch us, in effect, to learn more about what we want, which helps them disregard possible actions that might send everything to the devil, so to speak. Learning about human preferences enables computers to avoid hurting us when attempting to achieve their objectives. (Russell does not explain how a superintelligence still dumb enough to wipe us out of existence under the impression it's helping us should be trusted in the role of benevolent "panopticon," ever watching and learning about our preferences.)

Russell's retelling of the existential risk story about futuristic AI brings to mind the Czech playwright Karel Capek's universal robots, who were engineered for optimal work efficiency, deliberately absent other mind traits, like appreciating beauty, having a moral sense, and experiencing feelings and consciousness. The supposedly mindless automatons in his play *R.U.R.* got disgruntled somehow anyway, and sparked a robot revolution that wiped out virtually the entire human race. Capek's ending is no doubt why we remember his 1920s play. No one gets excited about the prospect of a souped-up Roomba superintelligently (yet mindlessly—ignore the contradiction) learning all about how best to vacuum, or clean the kitchen, or fix the car. Sure, it would be incredibly helpful, but it's not what we mean by superintelligence. We're excited about Ava. A superintelligence that isn't conscious or feeling or capable of diabolic aggression isn't really intelligent at all. Lacking common sense, too, it seems a poor candidate for our mythological imaginations. It's a calculator.

By tying human and machine intelligence together as, in essence, a game-theoretic quest to optimize objectives, Russell makes room for a seemingly "scientific" view of a computer mind, but only by severely restricting the possibilities of our own minds. This is an intelligence error once again. Human intelligence is various, still profoundly mysterious, and for all we know, effectively unbounded. By pulling down human intelligence, tying it to a definition more amenable to computation, current thinking about AI jettisons a richer understanding of mind. We are left with a simplified world.

Perhaps this world makes talk of a coming artificial general intelligence seem more reasonable (because "AGI" doesn't amount to so much), but it does so by jeopardizing interest in the project itself. We might as well retire the whole notion of superintelligence and begin a frankly more honest discussion about the very real possibility of globally destructive computer viruses, say, released by coders with manifestly ill-intent, mindlessly bringing down financial markets or hacking into and destroying data critical to the privacy of individuals or the security of countries. This is computation made effective by the discovery of vulnerability. It's the real world, not myth.

IN SUMMARY

We can summarize these positions about AI and people as follows. *Kurzweilians* (mythologists about AI, full-stop) wax mystical about machines after the Singularity having consciousness, emotions, motives, and vast intelligence. Ironically, they keep alive philosophical exploration by transferring Shakespearean themes to computation. (Computers will have rich spiritual experiences, and be great lovers, and so on.) We might call this the *Ex Machina* effect.

Russellians want to keep *Ex Machina* in movies, downsizing talk about superintelligence to more mathematically respectable ideas about general computation achieving "objectives." Unfortunately, Russellians tend to lump human beings into restricted definitions of intelligence, too. This reduces the perceived gap between human and machine, but only by reducing human possibility along with it. Russellians are thought leaders in a cultural trend, which I have called "the simplified world." As Jaron Lanier puts it, "A new generation has come of age with a reduced expectation of what a person can be, and of who each person might become."[18]

Kurzweilians and Russellians alike promulgate a technocentric view of the world that both simplifies views of people—in particular, with deflationary views of intelligence as computation—and expands views of technology, by promoting futurism about AI as science and not myth.

Focusing on bat suits instead of Bruce Wayne has gotten us into a lot of trouble. We see unlimited possibilities for machines, but a restricted horizon for ourselves. In fact, the future intelligence of machines is a scientific question, not a mythological one. If AI keeps following the same pattern of overperforming in the fake world of games or ad placement, we might end up, at the limit, with fantastically intrusive and dangerous idiots savants.

We will turn now to the science of AI, and it is here—in scientific enquiry itself—that the simplified world grows complex again, and mysterious. For, when we remove the constraint of our intelligence errors, the scales lift from our eyes, and a very formidable problem indeed presents itself.

Part II

THE PROBLEM OF INFERENCE

Chapter 8

• • •

DON'T CALCULATE, ANALYZE

AI is the quest for intelligence. Across the several chapters making up this part of the book, I hope to convince you that this quest faces significant obstacles, obstacles which we do not know how to surmount. To do so, we need to investigate the nature of intelligence itself. And there is no better place to begin our investigation than with a "strange and interesting young fellow," the amateur detective August Dupin, to whom we are introduced by the unnamed narrator in perhaps the world's first detective story, "The Murders in the Rue Morgue."[1]

ON SOLVING CRIMES

The narrator—who shares many traits with Edgar Allan Poe, the author—tells us early on that he's obsessed with the methods of thinking. He is curious about how human minds connect seemingly unrelated pieces of information with careful observation and reasoning—with inferences. What serendipity then, that the narrator finds himself lodging in an old house together with Dupin, spending all day around the brilliant detective.

Dupin, we soon learn, is not a normal guy. He's the sort of odd personality who possesses true originality. And indeed, he is odd. Dupin comes from an illustrious family but has been reduced to a state of near poverty, which he scarcely minds as he's constantly thinking, lost in ideas. When he does talk, he ruminates out loud. This could, of course, grow annoying. But the narrator treasures Dupin's "peculiar analytic ability." He says: "We passed the days reading, writing, or conversing, until warned by the clock of the advent of the true Darkness. Then we sallied forth into the streets arm in arm, continuing the topics of the day, or roaming far and wide until a late hour, seeking, amid the wild lights and shadows of the populous city, that infinity of mental excitement which quiet observation can afford."

Dupin is a prototype, an exemplar, like Sherlock Holmes. Like Holmes, he notices what the police, applying their "simple diligence and activity," somehow manage to miss.

One night, alone in the old house with Dupin, the narrator picks up the evening edition of the "Gazette des Tribunaux" and learns of murders in the Rue Morgue:

> "Extraordinary Murders.—This morning, about three o'clock, the inhabitants of the Quartier St. Roch were aroused from sleep by a succession of terrific shrieks, issuing, apparently, from the fourth story of a house in the Rue Morgue, known to be in the sole occupancy of one Madame L'Espanaye, and her daughter, Mademoiselle Camille L'Espanaye. After some delay, occasioned by a fruitless attempt to procure admission in the usual manner, the gateway was broken in with a crowbar, and eight or ten of the neighbors entered, accompanied by two gendarmes. By this time the cries had ceased; but, as the party rushed up the first flight of stairs, two or more rough voices, in angry contention, were distinguished, and seemed to proceed from the upper part of the house. As the second landing was reached, these sounds, also, had ceased, and everything remained perfectly quiet. The party spread themselves, and hurried from room to room. Upon arriving at a large back chamber in the fourth story, (the door of which, being found locked, with

the key inside, was forced open,) a spectacle presented itself which struck every one present not less with horror than with astonishment.

"The apartment was in the wildest disorder—the furniture broken and thrown about in all directions. There was only one bedstead; and from this the bed had been removed, and thrown into the middle of the floor. On a chair lay a razor, besmeared with blood. On the hearth were two or three long and thick tresses of grey human hair, also dabbled in blood, and seeming to have been pulled out by the roots. Upon the floor were found four Napoleons, an ear-ring of topaz, three large silver spoons, three smaller of metal d'Alger, and two bags, containing nearly four thousand francs in gold. The drawers of a bureau, which stood in one corner, were open, and had been, apparently, rifled, although many articles still remained in them. A small iron safe was discovered under the bed (not under the bedstead). It was open, with the key still in the door. It had no contents beyond a few old letters, and other papers of little consequence.

"Of Madame L'Espanaye no traces were here seen; but an unusual quantity of soot being observed in the fire-place, a search was made in the chimney, and (horrible to relate!) the corpse of the daughter, head downward, was dragged therefrom; it having been thus forced up the narrow aperture for a considerable distance. The body was quite warm. Upon examining it, many excoriations were perceived, no doubt occasioned by the violence with which it had been thrust up and disengaged. Upon the face were many severe scratches, and, upon the throat, dark bruises, and deep indentations of finger nails, as if the deceased had been throttled to death. . . .

"After a thorough investigation of every portion of the house, without farther discovery, the party made its way into a small paved yard in the rear of the building, where lay the corpse of

> the old lady, with her throat so entirely cut that, upon an attempt to raise her, the head fell off. The body, as well as the head, was fearfully mutilated—the former so much so as scarcely to retain any semblance of humanity.
>
> "To this horrible mystery there is not as yet, we believe, the slightest clew."[2]

The next day the *Gazette* publishes more details about the case. From the accounts of those giving evidence we can assemble the relevant information. The mother and daughter were well-to-do. Three days before the murders the mother had withdrawn a large quantity of money from the bank, in gold, which was found in open sight, untouched, on the floor after the murders. Also curious: a policeman among those first on the scene reported hearing two voices—one was obviously from a French-speaking man, and the other he could not recognize at all, calling it "hard, high, and very strange." He thought it from a foreigner, and possibly Spanish. Other witnesses would later describe the unintelligible voice as possibly Italian, Russian, or English.

Puzzling. Money—perhaps the most likely motive for murder—is left untouched in the house. The doors are locked from the inside. The body of the daughter is found up the chimney, lodged in with such force that it takes more than one person to pull her out. And, too, the voices. From the murderers, apparently, but although the police clearly heard two voices as they climbed the stairs to the home, only one is recognizable, the other reportedly a strange mixture of seeming gibberish. None of the witnesses are able to say exactly what is being said (if anything), in any language.

The police are flummoxed. Witness testimony only adds to the confusion. This is to say, all the clues, taken together, really point nowhere. The murders are a mystery, and that is precisely why our oddball amateur detective Dupin takes such a keen interest.

The narrator suggests that Dupin solves the case early on, by reading the account released by the police in the newspaper. The two, however, gain permission to visit the old house at the Rue Morgue while the crime scene is still intact. On the way back, Dupin stops in the office of another newspaper and places an ad in the lost and found section. Has someone in Paris, presumably a sailor, belonging to a Maltese vessel, lost their orangutan? The owner may call to claim it.

And here is the inference that cracks the case of the murders at the Rue Morgue: no human murdered the old woman and her daughter that night. The killer was not a human but a wild animal brought back from a jungle by a sailor, and kept in some nearby dwelling. In a frenzy after escaping his master, the orangutan leaped through the old house's window by swinging on the outside shutter, and then into the house, screaming and screeching and brandishing a straight-edge razor. Here's the murder weapon: the razor that beheads the old woman, and the sheer animal strength of the animal that crams the daughter feet first up the chimney.

The human voice heard by witnesses? The orangutan's owner. And the muffled and incomprehensible noises? The grunts of the animal.

THE GUESSWORK METHOD

But who would infer this from the facts of the case? Surely it is all right in front of everyone. In truth, Dupin just guessed. The police followed known methods until they led nowhere. Then, they started guessing too. The only difference was that Dupin's guess was the right one.

Poe begins "The Murders at the Rue Morgue" by ruminating about the nature of thinking. The fictional story of the crime begins first with nonfiction. He searches for the right words. Dupin's reasoning, he decides, is a triumph of analysis, in contrast to formulaic calculation. Calculation is connecting known dots; applying the rules of algebra, say. Analysis is making sense of the dots, making a leap or guess that explains them—and then, given some insight, using calculation to test it. Calculation has its limits: "But it is in matters beyond the limits of mere rule that the skill of the analyst is evinced." Rule-following isn't enough, but it is unclear what exactly else is involved. That Poe appreciates this mystery is evident in the declaration he makes at the outset of his story: "The mental features discoursed of as the analytical, are, in themselves, but little susceptible to analysis."[3]

The American scientist and philosopher Charles Sanders Peirce would read Poe's stories with fascination a few decades later. Peirce was also wondering how we think, how we reason about things. He managed even to capture Dupin's mental gymnastics in logical symbols. He didn't know how to automate the detective's insightful style of guessing, but he thought it was a central aspect of human thinking generally.

To Peirce, thinking isn't a calculation but a leap, a guess. Nothing is certain. We piece things together. We explain and revise. Peirce, living as he did in the nineteenth century, didn't know about digital computers. But he anticipated what would make AI a hard problem for everyone. It comes down, really, to this: Given that our own thinking is a puzzling series of guesswork, how can we hope to program it?

Eventually Peirce developed an entire explanatory framework for human reasoning. It was based on formal logic and its types, like deduction and induction.

And there was a third element, Peirce reasoned, that captured our guessing games. He called it "abduction." It is to this that we turn next.

Chapter 9

• • •

THE PUZZLE OF PEIRCE (AND PEIRCE'S PUZZLE)

For those familiar with his work, Charles Sanders Peirce is in a select group of truly original and important thinkers. The historian Joseph Brent, in his biography *C. S. Peirce: A Life,* called him "perhaps the most important mind the United States has ever produced." The philosopher Paul Weiss, writing in *The Dictionary of American Biography* in 1934, described Peirce as "the most original and versatile of American philosophers and America's greatest logician." The cultural historian and critic Lewis Mumford placed him in the company of iconoclastic geniuses like Roger Bacon and Leonardo Da Vinci. And when Noam Chomsky, the pioneering linguistics scientist at MIT, was asked in a 1976 interview about his influences, he said, "In relation to the questions we have been discussing [concerning the philosophy of language], the philosopher to whom I feel closest and whom I'm almost paraphrasing is Charles Sanders Peirce."[1]

BRILLIANT BUT ALONE

Like Albert Einstein, Peirce was left-handed and thought in pictures. He sketched out logical inferences in diagrams. In his later years, he wrote alone in his home, complaining that he was hungry and cold, too poor to afford fuel for the stove. His few friends worried about him and managed to get him a series of lectures at Harvard on the foundations of logic, in which he explained the types of logical inference with a framework that he thought undergirded the scientific method—a program for thinking clearly. Among the attendees was William James, the famous philosopher and early psychologist at Harvard, who later confessed he didn't understand the lectures completely—that the mathematics attached to Peirce's pictures and diagrams were beyond his ken. Apparently James wasn't alone in this; the lectures went largely unnoticed and appeared in book form only decades later.

Born into Victorian scientific culture in Cambridge, Massachusetts, in 1839, Peirce came from a well-to-do and overachieving family. His father was an eminent professor of mathematics at Harvard. A younger cousin would become a powerful senator, Henry Cabot Lodge. Peirce was classically educated, graduating in 1863 summa cum laude from Harvard University's Lawrence Scientific School. He spent thirty years as a research scientist with the US Coast and Geodetic Survey on topological studies of the earth's surface using precise measurements of gravity intensity. He was an amateur chemist, a prestigious lecturer on logic at Johns Hopkins University, and the first American delegate to any international scientific association. He was a scientist, a logician, a philosopher, a writer, a prolific book reviewer for the *Nation*, and more. Peirce scholar Max H. Fisch, who spent decades researching Peirce's life and work, offers this suitably grand judgment of his many achievements:

> Who is the most original and the most versatile intellect that the Americas have so far produced? The answer "Charles S. Peirce" is uncontested, because any second would be so far behind as not to be worth nominating. Mathematician, astronomer, chemist, geodesist, surveyor, cartographer, metrologist,

> spectroscopist, engineer, inventor; psychologist, philologist, lexicographer, historian of science, mathematical economist, lifelong student of medicine; book reviewer, dramatist, actor, short story writer; phenomenologist, semiotician, logician, rhetorician, metaphysician. . . . He is the only system-building philosopher in the Americas who has been both competent and productive in logic, in mathematics, and in a wide range of the sciences. If he has had any equals in that respect in the entire history of philosophy, they do not number more than two.[2]

Yet Peirce died an outcast, largely forgotten. Forgotten geniuses are common enough in history that we occasionally rediscover them, as with Tesla. But arguably more than Tesla—who, after all, achieved a kind of posthumous fame as Elon Musk's choice of inspiration to name an electric car company after—Peirce stands as an important thinker who has been mostly written out of the history books. His work is most appreciated in philosophy, where he is known as the founder of the philosophical school known as pragmatism.

His early work on computing has been all but forgotten. Scholars still plumb his voluminous writings on the nature of logic, but the subject is arcane, too difficult to join up with mainstream discussion. Thus, even as some who have understood the significance and scope of his thoughts on the nature of logic have compared him to Aristotle, a discussion of Peirce's ideas today requires in most circles a biographical sketch—and an explanation, even an apology.

PHYSICS, PHILOSOPHY, AND PERSONALITY

Why was Peirce forgotten? His personal life gives us a clue: he irritated nearly everyone. William James remained a close and lifelong friend. But even James came away from his first encounter with Peirce, when both were students at Harvard, with a mixed impression, as he wrote to his family: "there is a son of Prof. Peirce, whom I suspect to be a very smart 'fellow' with a great deal of character, pretty independent and violent though."[3] Sympathetically, James later referred to Peirce as "that strange and unruly being."[4]

Peirce's prickly personality and unconcern with contemporary mores attracted endless trouble for him personally and professionally. He often offended Victorian socialites in New England (including his family), who justifiably shunned him; Harvard refused to offer him a professorship because of a known infidelity in his marriage; the Coast Survey of the United States government, where he worked for decades, ultimately fired him for failing to deliver reports on time and for losing expensive equipment while touring Europe. So, too, was he dismissed by Johns Hopkins, after unspecified reports of unbecoming conduct. Today, we would say that he didn't fit in—a perfect stereotype of the misunderstood genius. He was constitutionally incapable of playing by the rules.[5]

Peirce's personal scandals and idiosyncrasies help explain why, for instance, records of his personal life—volumes of documents—remained sealed in Harvard's Houghton Library until 1956, forty-two years after his death. His scientific and philosophical papers—including many of immense interest to computer science and especially AI—lay relatively untouched, and unpublished, in Harvard's archives for want of a "few thousand dollars" to "guarantee the initial expense of publication," as Lewis Mumford later put it.[6] Not understanding his ideas and their significance, and unwilling to invite scandal, many of those who knew Peirce never saw his reputation restored or much of his life's work published. Peirce himself died in obscurity, survived by his equally enigmatic second wife, Juliette, a French woman with her own somewhat checkered history. Fittingly, Peirce's family sometimes described Juliette as an outcast, or a "gypsy."

Only much, much later—and to some extent not even still today—have we come to understand the enormity of Peirce's contribution to mathematics and especially logic. Of utmost significance is his thinking about logical inference, and in particular, to move to the centerpiece of his life's work, his exploration of the depth and mystery of what he called *abductive inference,* a kind of explanatory guess that, he realized, undergirds most of our thinking.

Peirce noted that abductive reasoning had been left out of accounts of logical reasoning going back to Aristotle. It also didn't fit into the usual logical framework assumed in mathematics or logic courses. He saw abduction as a missing logical piece, which raised fundamental questions about automation and intelligence. Had he known about AI, he likely would have seen what is today too often missed: that the problem of abductive inference confronts AI with its central, still entirely unsolved, challenge.

THE PUZZLE OF INFERENCE

Edgar Allan Poe's narrator groped for words to describe what Peirce would later write volumes about—abductive inference. But abductive inference is a kind of inference. What is inference? A noun, for one. The verb form is "infer," which speaks to actions. Etymologically, to infer means to "bring about," from Latin "in," into, and "ferre," bring. The *Oxford English Dictionary* tells us it's something we do cognitively, with our minds: "to reach an opinion or decide that something is true on the basis of information that is available."

Unfortunately, the *OED* also tells us that "deduce" is a synonym of infer, which is unhelpful (deducing is only one way of inferring).

The *OED* also offers a few usage examples that highlight the generality of the word "infer" in common parlance:

> *To infer something (from something):* Much of the meaning must be inferred from the context. Readers are left to infer the killer's motives.

To infer that: It is reasonable to infer that the government knew about these deals.

Inference is to bring about a new thought, which in logic amounts to drawing a conclusion, and more generally involves using what we already know, and what we see or observe, to update prior beliefs. We might infer the killer's motives (to borrow from the *OED*) using what we already know and what we've read in the papers (like Dupin).

Inference is also a leap of sorts, deemed reasonable, as when we infer that "the government knew about these deals"—again, on the basis of whatever prior knowledge we have (like public or shared knowledge) as well as reading (observing) some breaking story or stories.

Inference is a basic cognitive act for intelligent minds. If a cognitive agent (a person, an AI system) is not intelligent, it will infer badly. But any system that infers at all must have some basic intelligence, because the very act of using what is known and what is observed to update beliefs is inescapably tied up with what we mean by intelligence. If an AI system is not inferring at all, it doesn't really deserve to be called AI. (Although we might say that even a system that tags pictures of cats is inferring that what it "sees" is a cat—so the bar can be quite low.)

It's impossible to get a joke, discover a new vaccine, solve a murder like Dupin does, or merely keep up with sundry happenings and communications in the world without some inference capability or other. We know a lot of things, sure, but only inference gets us to new knowledge (or belief). We know that the sun will rise tomorrow, so we don't need to infer it. Likewise we don't bother inferring that our hand is still attached to our arm. This is knowledge we already have, a set of beliefs we've already formed. But our knowledge is always changing and getting updated. If it's mysteriously dark outside too early, we might infer a solar eclipse, or maybe that a large dust storm is blotting out the sun to the west, or maybe that there's a nuclear holocaust. It depends—what do we currently know? What makes most sense of what we see?

In a general sense, we're always inferring—it's like a condition of being awake. I might walk into the kitchen, discover a half-empty can of Pepsi, and infer that my sister left it there, as she drinks Pepsi and is visiting. On the other hand, there are workers here redoing the countertops, and I also noticed that one of them was drinking a Pepsi earlier. For that matter, I was drinking a Pepsi earlier and left it unfinished on the porch, so perhaps my spouse brought it in. We end up guessing an explanation that makes sense, given what we know and the context we're in. This is "real-time" inference, since we're drawing conclusions as we walk into the room. Real-world circumstances are always changing, so real-time inference is common. After all, we think in time. A computer program that can solve a problem after ten billion years is not intelligent at all, and neither is one that, in real time, walks into a wall.

The provisional nature of many inferences means that initial ones can be wrong, especially if arrived at too hastily. If I come in to the office late, the boss might infer that I'm not taking things seriously, when in fact the traffic was backed up due to an accident. In other words, the boss draws a conclusion based on some preformed impression or prejudice about me. People in day-to-day conversation use the word inference in this sense, referring to a too-hasty jump to an unwarranted conclusion: "Oh, that's Suzy, she's inferring all sorts of crazy things about you after what you said last night." And it is true that, technically, Suzy is making inferences, but the sense here is that they are biased ones, with Suzy too ready to make unfair assumptions (perhaps because she's in a bad mood or doesn't like you).

In a more particular sense, inference entered the mathematics lexicon long ago, and has featured more recently in discussions about computation and AI. In this setting, "real-time inference" might refer to a robot navigating a dynamic environment, like a busy street.

"Probabilistic inference" draws conclusions from statistical data, with obvious application to data-centric approaches to AI.

Once upon a time, AI scientists struggled mightily with a precondition of inference, the intelligent use of what we already know—the question of "knowledge." Systems that don't know anything can't infer much, either. So early researchers tried to code knowledge into AI systems, to help them make sense of their sensor or text input. It was discovered (the hard way, by repeated failure) that AI systems with large knowledge repositories of facts and rules still had to use the knowledge in context, to draw relevant conclusions. This "using" of knowledge is what makes inference so hard. Which bit of knowledge is relevant in the haystack of my memory, applied to the dynamically changing world around me?

The ability to determine which bits of knowledge are relevant is not a computational skill. Poe insists that in the realm of the "analytic," human insights are not arrived at by formula; they are "matters beyond the limits of mere rule" or calculation. Indeed, Dupin seems to arrive at his explanation for the killings—the orangutan—by a kind of serendipitous guess, which he later verifies by meeting the owner of the missing animal. So: was he just guessing? In an important sense, he was. But this doesn't cancel it as an inference. It makes it an important kind.

MORE ON TURING

In his seminal 1950 paper "Computing Machinery and Intelligence" Turing dismissed questions about machines actually thinking, poking fun at his own title, claiming that "thinking" is hopelessly unscientific and subjective. Talking about computers thinking is like talking about submarines swimming. To refer to "swimming" is already to anthropomorphize. Dolphins swim, but submarines don't. Turing thought the use of the word thinking was like this, too. If a computer played chess, who could say whether it was thinking or just calculating?

Turing was interested in a fully programmable mind. He therefore disposed of his original distinction between insight and ingenuity by pulling insight—whatever it is—into the sphere of computation. By so doing, he made the question of AI fully testable. The thesis was radical even by his own earlier standards, but we will not begrudge him for that because it laid the groundwork for AI researchers later in the decade to begin work without philosophical worries holding up progress.

Unfortunately, exactly how computational inference could be—or would become—like human inference was never adequately addressed. The field didn't start with a theory of inference, which would have provided a blueprint for later development (or an impossibility proof). For AI researchers to lack a theory of inference is like nuclear engineers beginning work on the nuclear bomb without first working out the details of fission reactions. Knowledge of Einstein's equation is not enough, clearly. And knowledge of computational theory by AI enthusiasts isn't, either—because the very question confronting scientists working on AI is how computation can be converted into the proper range and types of inference exhibited by minds. The question had to be asked directly. By ignoring or skirting it, the field made false hopes, dead-end paths, and wasted time inevitable.

For there is much to consider. Take, for instance, the many inferences found in the history of science. Scientists frame hypotheses, then test them. But the hypotheses aren't arrived at mechanically; notoriously, they sometimes pop into scientists' heads (typically after mastery of the field). Like Turing once did, students of scientific discovery tend to push such intellectual leaps outside the formalities of scientific practice, and so the central act of intelligence "rides along for free," unanalyzed itself. But such hypotheses are genuine acts of mind, central to all science, and often not explainable by pointing to data or evidence or anything obvious or programmable.[7]

When Copernicus posited that the earth revolved around the sun and not vice versa, he ignored mountains of evidence and data accumulated over the centuries by astronomers working with the older, Ptolemaic model. He redrew everything with the sun at the center, and worked out a usable heliocentric model. Importantly, the initial Copernican model was actually less predictive despite its being correct. It was initially only a framework that, if completed, could offer elegant explanations to replace the increasingly convoluted ones, such as planetary retrograde motion, plaguing the Ptolemaic model. Only by first ignoring all the data or reconceptualizing it could Copernicus reject the geocentric model and infer a radical new structure to the solar system. (And note that this raises a question: How would "big data" have helped? The data was all fit to the wrong model.)

The Copernican leap that launched the Scientific Revolution could better be described as an inspired guess. The same could be said of Kepler's choice of an ellipse to describe planetary motion, because a vast number of (technically, infinite) geometric shapes can be fit to planetary orbits (perhaps excluding transcendental ones, like sine waves). The ellipse wasn't simpler than all the others—this wasn't an Occam's razor explanation. What Kepler literally conjectured was an explanation that to him "felt right."

The fact that conjectures lead to discoveries doesn't fit with mechanical accounts of science; to the contrary, it contradicts them. But detective work, scientific discovery, innovation, and common sense are all workings of the mind; they are all inferences that AI scientists in search of generally intelligent machines must somehow account for.

As you can see, cognitive modeling—building a computer to think, to infer—is puzzling. AI researchers (at least for now) should be most concerned with inference in its everyday context. Why? Because the vast majority of inferences we make are seemingly mundane, like all the multifarious leaps and guesses made in the course of ordinary conversation. Unfortunately for researchers in AI, even mundane inferences are not simple to program. The Turing test, for instance, is hard essentially because understanding natural language requires lots of commonsense inferences, which are neither logically certain nor (often) highly probable. It requires, in other words, lots of abductions.

We typically don't even notice such inferences, which is good: if we did notice them, we'd tend to get stuck in solipsistic loops, chasing around our thoughts. This brings us back to Peirce, and more specifically, it brings us to the tripartite inference framework undergirding intelligence: deduction, induction, and abduction.

Chapter 10

• • •

PROBLEMS WITH DEDUCTION AND INDUCTION

Throughout most of intellectual history, inference has been synonymous with deduction. Aristotle studied a simple form of deduction known as the syllogism—two statements known or believed already true, leading to a third, the conclusion. Aristotle developed an early form of logic using syllogisms to analyze arguments made by himself and others, and to lay a foundation for correct reasoning. In his tradition, intelligence must conform to known deductive rules.

This makes sense. We should not be swayed, for instance, by someone arguing that Ray Charles is God because *God is love,* and *Love is blind* (and so is Ray Charles). The argument is fallacious—it breaks rules of deductive reasoning. Writing all this down precisely has been the tradition of deductive logic. Aristotle also explored how deductive rules relate to so-called practical reasoning—for instance, when an intelligent agent formulates a plan to achieve a goal whose steps can be analyzed logically. (The plan might be provably "correct" yet fail in the execution—still, it's a start).

Logical (correct) reasoning and planning are important subfields in AI and, almost since its inception, classic AI has explored approaches to reasoning and planning using symbolic logic such as deduction. An AI system can implement a syllogism, for instance, and also a planning algorithm (rules of the form: $\{A, B, C, \ldots\} \rightarrow G$, where A, B, and C are actions to be taken and G is the desired goal). There have been no major breakthroughs toward artificial general intelligence using such methods, but even modern AI scientists like Stuart Russell continue to insist that symbolic logic will be an important component of any eventual artificial general intelligence system—for intelligence is, among other things, about reasoning and planning.

Aristotle thus kicked off formal studies of inference thousands of years ago. A few decades ago, he also helped kick off work on AI. Symbolic reasoning using rules from deduction ties intelligence specifically to knowledge, a prerequisite for common sense, which is still missing almost entirely from AI systems. Early AI pioneer John McCarthy (a founder of the field, at the Dartmouth Conference in 1956) realized this early on, launching a sustained effort at the development of knowledge-based systems—systems that rely on computer-representable statements about the world to reason and act. All the old knowledge-based systems ran into defeating, if instructive, problems. Some of these problems, perhaps, may be revisited with hope of progress. Others, however, seem fundamental. In particular, they are limitations inherent in rule-based reasoning itself. Deductive logic is precise because it gives us certainty. As we might expect, certainty is a high bar for the real world, where artificial general intelligence systems (and people) must prove their intelligence.

DEDUCTION: HOW NEVER TO BE WRONG

Logicians (and computer scientists) analyze deductive inference in systems of statements that can be true or false. By convention, all statements written before the last in a series are called premises. The last statement follows from the premises; it's called the conclusion. The statements and the conclusion together are known as an argument.

A good deductive argument is a "sure bet," because its conclusion is necessarily true. Here's one:

If it's raining, the streets are wet.

It is, in fact, raining.

Therefore, the streets are wet.

The conclusion is the inference we should draw from the two premises. (In essence, it answers the question: Knowing nothing else, what follows from the premises?) The rule used to infer the conclusion is called valid if the conclusion must be true whenever the premises are true. Validity is a "trustworthy" stamp for the rule used; the rule will always preserve truth, whenever our premises (or prior beliefs) are true. Thus, the above example is valid. It uses one of the oldest deductive rules discovered, still referred to in Latin: *modus ponens*. In quasi-symbolic form:

If P, then Q

P

Therefore Q

And in fully analyzable (computable) format, we have its logical form:

$P \rightarrow Q$

P

———

Q

Here, the connector "→" has a specific meaning, or semantics, which determines the truth values for P and Q. In deductive logic the rule is called a material conditional, and guarantees that Q follows from the truth of P and the rule $P \rightarrow Q$. (The range of true or false possibilities is given by a truth table, shown later.)

Now consider some modifications to our argument about rain and streets. In particular, what if it's not raining? In that case, the rule doesn't "fire." Nothing follows. But the argument form is still valid. It's still true that if it's raining, then the streets are wet. If in fact it *is* raining, then the argument becomes "sound" (and not just valid). Soundness is truth—real truth, as opposed to the conditional truth of validity. Soundness tells us that the premises really are "true." Soundness guarantees that intelligent agents using deductive inference will infer truths from prior truths. Validity, on the other hand, guarantees only that whatever the intelligent agent believes, its inferences will be formally correct (even if reasoning about lies or falsehoods). In fact, deductive arguments that are valid but not sound can introduce all sorts of silliness into deductive reasoning. For instance:

> If it's raining, then pigs will fly.
>
> It's raining.
>
> Therefore, pigs will fly.

A silly argument, but perfectly valid, because it again uses *modus ponens*, the mode of reasoning from a hypothetical proposition. The first premise is of course false. The second premise might be false, too, if it is not actually raining. Even if it is raining, however, we can't rely on the first premise, because there's no connection between rain and the flight of pigs—and anyway, pigs don't fly, regardless of the weather or anything else. The argument is valid, but not sound—and perfectly useless.

Here's a sound deduction:

> All men are mortal.
>
> Socrates is a man.
>
> Therefore Socrates is mortal.

How can it be wrong? It can't. The conclusion always follows with one hundred percent certainty. Deduction supplies a template for "perfect" and precise thinking for humans and machines, and primarily for this reason it has been investigated extensively in mathematics and the sciences, and used successfully in several important applications in the field of AI. Early on, for instance, deduction-based AI systems were able to automatically prove real (not "toy") theorems in mathematics. A computer program called Logic Theorist, the brainchild of AI pioneers Alan Newell, Herb Simon, and Cliff Shaw, proved interesting logical theorems as early as 1956, using the foundational twentieth-century work on logic, Bertrand Russell and Alfred North Whitehead's *Principia Mathematica*. Automated reasoning systems using deduction have also been applied to circuit design for computer motherboards, and to the task of software and hardware verification, ensuring software doesn't contain bugs or contradictions.[1] In such cases, the deductive approach is easier and more effective than modern AI methods using statistics and learning. As early researchers in AI knew, too, our knowledge is often expressed symbolically (as in the rain example above), so deduction makes sense; it's an obvious choice. Unfortunately, there are well-known problems in extending deductive inference to general intelligence.

KNOWLEDGE PROBLEMS

Over the years, many problems with deduction have been discovered. Perhaps the most damning: deduction never adds knowledge. If I know that people are mortal (they die) and that such-and-such is a person, I already know that such-and-such will die. The deduction simply confirms what a rational person should conclude from the premises given, which in a simple syllogism is easy to see because the "knowledge" is already contained in the statements. The conclusion just makes it explicit.

Deduction is extraordinary useful as a defense against someone inferring wild or incorrect conclusions from a set of statements—say, by insisting that based on the premises of human mortality and Socrates's being a human, we should conclude that Alpha Centauri is made of cheese. Deduction gives rational agents a template for "staying on track," which is clearly a good first step for any AI system we hope will make intelligent inferences. But we don't get very far using only deduction. For example, in response to Copernicus's theory that the earth revolves around the sun and not vice versa, old-school Ptolemaic astronomers might employ a deductive counterattack:

> If the heavens were created by God, Earth would be at the center of the heavens.
>
> The heavens were created by God.
>
> Therefore, Earth is at the center of the heavens.

The argument is valid, but again, this tells us only that if the premises are in fact true, then the conclusion necessarily follows. All the heavy lifting is in empirical questions about the veracity of the premises. We got this "for free," so to speak, with our inquiry into Socrates's mortality, since generally we all agree that people die (even if they go to heaven later). But the generalization that any heavens created by a divinity would feature our own planet at the center seems as debatable as any other aesthetic or scriptural assertion. We might insist on a variant interpretation of scripture (Galileo famously remarked that God tells us how to go to heaven, not how the heavens go). Or we might, especially if we were atheists or scientific materialists, reject the truth of the second premise out of hand.

Deduction is therefore useless in the pursuit of new knowledge, and it only clears up disputed beliefs if bona fide errors in reasoning are made. Famously, conspiracy theorists might never make deductive reasoning mistakes—it's just that they adopt premises as true that others find dubious or just crazy.

In other words, any intelligent system will require other types of inference to zero in on true (and useful) beliefs in the first place. A deductive certainty in inferred conclusions isn't enough.

RELEVANCE PROBLEMS

Deduction has other limitations that make it unsuitable as a strategy for engineering general intelligence. One particularly damning one involves considerations of relevance. The premise *If it's raining, then pigs will fly* is untrue, because pigs don't fly, but it's also a fantastically bad example of saying something relevant. Rain has nothing to do with pigs flying. On the other hand, planes do fly, but the premise *If it's raining, planes fly* is irrelevant, too. It might be true (at least some of the time), but from the fact that it's raining we shouldn't hold beliefs about planes in the air. The statement again ignores considerations of relevance.

Part of the problem here is causation: rain doesn't cause planes to fly (though it might in some circumstances keep them on the ground). Here it depends on how we want to use the knowledge. *If the thermometer is in the red, it's hot outside* is true. But if we want to infer a likely explanation for a heat wave, the thermometer isn't any help. The statement is true but irrelevant. *If the rooster crows, the sun is coming up* is also true, but if we were to ask an artificial general intelligence system why the sun rose and it offered up the rooster, we'd be reluctant to attribute much intelligence to it.

Consider this example, taken from the philosopher of science Wesley Salmon:

> All males who take birth control pills regularly do not get pregnant.
>
> A man takes his wife's birth control pills regularly.
>
> Therefore, the man does not get pregnant.[2]

In fact, this is a perfectly sound deductive argument: it uses *modus ponens* with true premises. But the man's avoiding pregnancy has nothing to do with the reasons given. They're irrelevant, because men don't get pregnant anyway. The argument explains nothing. We can imagine a robot armed with a vast database of facts and rules reasoning in this way, using deduction. Nothing is really wrong, per se, but the robot doesn't understand anything—it doesn't know what's relevant and what's silly.

Consider a subtler example:

> Anyone who eats an ounce of arsenic dies within twenty-four hours.
>
> Jones ate an ounce of arsenic at time *t*.
>
> Jones died within twenty-four hours of *t*.

This is a perfectly fine deductive argument, but it would not explain Jones's death if, for instance, Jones ate the arsenic at time *t* and died from a traffic accident (perhaps racing to the hospital) before he expired from the poisoning. Here again, the argument is good deduction, but irrelevant. It tells us nothing. It's even misleading. Relevance, in other words, often presupposes knowledge of causation, where some event actually brings about a result, or makes something happen.

Another reason deduction falls endlessly victim to relevance problems is that there are, invariably, many possible causes for the occurrence of something in our day-to-day experience (and in science). Accidents like aircraft crashes, for instance, can typically be analyzed by pointing to proximate (close by) and distal (farther away) causes, together explaining the disaster. Take the recent Boeing tragedies. After two crashes of Boeing 737 Max planes occurred in the span of six months in 2018, investigators discovered a software glitch in an anti-stall system, the Maneuvering Characteristics Augmentation System (MCAT). A redesign of the older Boeing 737–800 had enabled larger engines to be fitted, but only by placing them forward of and slightly above the wings. This resulted in steeper climb rates on takeoff, which could induce, under certain conditions, a stall. Stalling is bad—potentially catastrophic—so the MCAT was fitted to the new Max to push the nose down when necessary to avert a stall. Unfortunately, the nose-down correction of the Max could send the plane hurtling toward the ground. The MCAT did just this, taking control away from the pilots in two tragedies resulting in the death of 157 people in Indonesia and 189 in Ethiopia.

Subsequent investigation revealed flaws in the software controlling the MCAT, so a proximal cause was identified. But the ensuing investigation also highlighted Boeing's zeal in pushing the Max into service to compete with fuel-saving aircraft offered by rival Airbus—pointing to a background or perhaps distal cause. It was also discovered that pilots of the new Max received inadequate training. This was surely not helped by Boeing's marketing pitch for its redesigned aircraft, claiming that the Max would not require expensive retraining of pilots already trained on the 737–800. Thus, the tragic crashes can be attributed to multiple causes. Inferring why Boeing's 737 Max crashed involves considering a number of possible causes, and perhaps no single cause by itself fully accounts for the catastrophes.

Deduction can't speak to these real-world scenarios. By requiring that inferences must certainly be true, deduction invariably misses what might be true, in contexts where relevance is determined by a mix of factors that aren't necessary but still are operative in certain situations. In Plato's universe of unchanging forms, triangles must have three sides, and some things are True with a capital T. In messy experience, few things we witness or analyze are like triangles. They're like the Boeing 737 Max—or an ordinary conversation (as we'll see). Intelligence—whatever it is—is more than deductions. We are cognitive systems ourselves, and it's clear that we're not only deductive systems. Successful human-level AI, this suggests, can't be wholly deductive either.

After the failure of what critics dubbed "good old-fashioned artificial intelligence," which dominated AI before the modern era (up through the 1990s), AI scientists abandoned deductive approaches to inference en masse. Indeed, many younger readers might find it strange that something like "rules" and deductive approaches to AI were ever taken seriously by practitioners in the field. They were. But the devastating limitations to deductive inference eventually spelled doom for the approach. And, as the web exploded, the volumes of data available for so-called shallow or statistical methods made deductive or rule-based systems seem less useful, and clunky. A new paradigm—a different type of inference—came to prominence in serious work on AI. It's called induction, and we turn to it next.

THE POWER AND LIMITS OF INDUCTION

Induction means acquiring knowledge from experience. Experience is typically construed as observations—seeing things—although it can also come from any of our five senses. (Touching a hot stove is an example of tactile induction.) The general form of induction, unlike deduction, is from particular observations to general hypotheses. The induced hypothesis covers—that is, explains—an observation. The primary mechanism of induction is enumeration: it is hard to induce the features of a population of, say, birds (to use a famous example) without first observing many examples of birds. The centrality of enumeration plays a central role in all versions of induction, and will be important to understanding its nature and limitations.

Induction is powerful not just because it helps organize the world of things into categories via hypotheses (all these objects of *X* have property *Y*); it also confers predictive power on agents who use it.[3] If every time a major league baseball game ends the streets downtown are packed with people, I might infer that the next time a game wraps up they will be again—this is a prediction. Induction captures the everyday idea that we gain the ability to explain and predict by observing happenings in the world. Many of our expectations are induction-based. If someone were to move the doorknob on your front door ten centimeters to the left, you'd likely miss it when reaching for it. You have an implicit theory—that is, a hypothesis—of where the doorknob is, based on many prior examples of seeing it and grabbing it.

Induction has other virtues. It's synthetic, for one, to borrow Kant's phrase; it adds knowledge. I might look online for when traffic is at its peak at the corner of Third and Main, but if I work at Third and Main I can look out the window. The latter is firsthand observation that facilitates my inductive inferences, forming expectations and plans for when I should leave. Unfortunately, the powerful flexibility of induction (tied to our senses) means also that it can't be provable, or guaranteed true, like deduction. Knowledge gleaned from observations is always provisional. Why? Because the world changes. The future could falsify my inductive hypotheses. My car might have started without a hitch a thousand times. Tomorrow morning (when I'm late setting out to a meeting—Murphy's Law), it might not. This is induction. Change comes (or it doesn't, alas), and prior observation alone won't tell us how or when.

The strength of induction, though, lies in the fact that intelligence is importantly tied to looking at the world around us. Modern science would be impossible without allegiance to induction as a means to knowledge through experience.

Consider again enumeration. In its simplest form, induction requires only the enumeration of prior observations to arrive at a general conclusion or rule (or law). Here's an argument:

> *N* swans observed have been white [where *N* is some large number].
>
> Therefore, all swans are white.

Or:

All life we've ever seen has been carbon-based.

Therefore, all life is carbon-based.

As these examples suggests, simple enumeration (that is, counting) of the features or properties of something often forms the basis for our claims to knowledge about the thing as a type. Thus, swans are just those birds that are white; life is just that phenomenon arising from carbon. In science (and in life) we also find it helpful to tell a story about why swans might be white, or life might be carbon-based, but strictly speaking, explanations answering such *why* questions are outside the scope of induction, enumerative or otherwise.

It's the simplicity of induction, though, that gives it such utility as a type of inference. The more I observe some property in some object, the more confidence I have that the property is part and parcel of the object. If I keep sampling balls from a bag, and they are always white, at some point I'll become confident in a generalization like *All the balls in this bag are white*. But, again, if I haven't sampled every single ball, it's always possible that my inductive inference will be wrong. Induction is useful but not certain knowledge.

Here's another type of inductive generalization:

The proportion Q of a sample of the population has property *P*.

Therefore, the proportion Q of the population has property *P*.

Induction from sample to population is quite common in scientific investigations, and sophisticated statistical techniques have been developed over the years to help make these generalizations as strong and error-free as possible given available observational evidence. Intuitively, too, inductive generalizations make sense: if I observe 75 white balls and 25 black balls in some sample, then, absent other evidence, I should expect 750 white balls in the population of 1,000. The inference seems right; it's just not certain.

Random sampling is also based on a generalization from observations. Try it yourself: flip a coin repeatedly and count the heads and tails that come up. This is a random sampling (since you can't skew the coin flip if it's a fair coin). You might flip two or three heads in a row. Very improbably, you might even flip five of all heads or all tails. But given a large enough sample, you can generalize that the odds of a coin landing either heads or tails are fifty-fifty. Hence, the inductive generalization is *The coin will land heads five hundred out of a thousand times,* which gets you close enough. (The law of large numbers tells us that given a large enough sample, the probability will approach the actual probability: across a million coin flips, the fifty-fifty split will be quite close). Here's another popular example of statistical generalization using induction:

> Seventy-three percent of randomly sampled voters are for Candidate *X*.
>
> Therefore, Candidate *X* will get about seventy-three percent of the votes.

Candidate *X* might become embroiled in a scandal before the election, invalidating the inductive inference. But again, absent more knowledge, we can reason this way and draw conclusions about what we expect to happen.

Modern AI is based on statistical analysis and so relies on an inductive framework, which is useful for many commercial applications. For example, AI can offer recommendations—a type of prediction based on past observation. Here's another example familiar to anyone with a content feed:

> Seventy-five percent of the news User *X* reads is conservative political commentary on website *C*.
>
> Therefore, User *X* will want this next piece of news on *C*.

User *X* might also like to read the occasional article from the *New Republic.* Unfortunately, the system inferring *X*'s preferences using induction is likely to filter it out. This is one obvious drawback to relying on inductive generalizations from observation—they are a surrogate for deeper knowledge (and even worse, they tend to expect the future to look like the past).

The eighteenth-century philosopher David Hume, who first pointed out the limits of induction, gave philosophers and scientists what is now known as the problem of induction. As Hume put it, relying on induction requires us to believe that "instances of which we have had no experience resemble those of which we have had experience." In other words, the general inductive rule we apply requires extension to unseen examples, and there is no guarantee that it will hold. Unlike deduction, there is nothing in the structure of induction that provides us logical certainty. It just works out that the world has certain characteristics, and we can examine the world and tease out the knowledge that (we think) we have about it.[4]

The problem of induction may seem like an armchair worry that philosophers like to indulge, but in fact the limits of inductive inference raise constant problems for scientists in their quest for true theories. Examples are everywhere. We used to eat egg whites because nutrition science warned us of the evils of saturated fats, found in the yolks. Fast forward a few decades, and discover that nutrition scientists now encourage us to eat the eggs, yolk and all. They help burn fat and raise your mood; they even protect against heart conditions (the very worry about them a few decades ago). In a very real sense, we can blame induction for these embarrassing about-faces. They happen because our observations and tests are never complete. Correlations might suggest an underlying cause we can rely on (a bit of real knowledge), but we might have missed something when testing and observing what affects what. The correlation might be spurious, or accidental. We might have been looking for the wrong thing. The sample size might be too small or unrepresentative for reasons that only become apparent later. It's a common problem, and at root it's just the specter of induction and its limits—the philosophers weren't wasting our time, after all.

At root, all induction is based on enumeration. This might (or should) seem suspiciously simple: Is it possible that to come up with theories about the world, we can just count examples? In an important sense, yes. A single experience doesn't license an inductive inference. If I see an orangutan and know what one looks like, I can classify it. But not knowing yet what the animal is, I'll need to observe many of them before I know whether the animal I saw is a freakish chimpanzee or the baby of an adult Big Foot. As Hume put it, we need to see "constant correlations" to infer causes, and we need to see enumerated examples to infer categories or types. (This is exactly how machine learning works, as we'll see.)

Of course, inductive reasoning gets more complicated: statistical inferences in fields like economics or the social sciences are also inductive, but one has to know a lot about probability theory (and economics and social science) to understand them. And new inductive inferences in the sciences inevitably build on older ones scientists now believe to be solid and true. (So we have to know about all those other theories, too.) But at root, induction simply generalizes from looking at examples. When the generalizations can be explained with some story, some cause or set of causes that make it so, then we're confident new knowledge has been acquired, even if it's not necessarily true, like deduction. It is supported by observation and testing.

Hume's critique of induction was primarily a critique of causation. Induction doesn't require knowledge about causes (in that case it wouldn't be enumerative). If we know, for instance, that the color of bird feathers is determined in part by characteristics of habitat, then even if all the swans in England are white we might expect black feathers on swans in different habitats. But, in the absence of theory, induction can tell us this only if we fly around the world and keep observing swans where they live. Hypotheses that cite specific causes are the goal of observation, but unfortunately the logical resources of induction are inadequate to supply them. Additional inferences are required (and here, deduction can help, but only partially).

The point is: induction properly understood in a logical framework of inference is, while necessary and common, quite limited. It is often misunderstood, too, which contributes to a general overconfidence that induction ensures "scientific" and solidly empirical knowledge, ridding us of fanciful speculation. Our detective hero Sherlock Holmes sometimes explains his method as painstaking inductions, simple and clear observations uncluttered by opinions and ideas and beliefs getting in his way. He assures a puzzled and amazed Watson that he just "observes things carefully." Holmes knows the value of simple observation—the simpler the better—because what we think we know can prevent us from seeing anything new. But this is only part of the intelligence story. We have to understand the significance of what we observe. Holmes, like Dupin, solved crimes by piecing together observations in a novel way. All the devil's details are in the novelty, which isn't induction at all.

Inductive inference presents us with another unavoidable danger, made memorable by Hume's critical eye: newly discovered facts can surprise us. In dynamic environments like everyday life, observation is open-ended. Future observations can reveal what was previously hidden and unknown to us—surprise! And our very confidence in inductive inference can make it harder to watch out for its inevitable shortcomings and failures. This brings us to holiday celebrations, or at least to Bertrand Russell's "inductivist turkey."

RUSSELL'S TURKEY

Bertrand Russell is one of the most famous philosophers and public intellectuals of the twentieth century. A logician, mathematician, and social activist, he once spent six months in prison for protesting

Britain's entry into World War I. Later, in the 1950s, he protested nuclear weapons proliferation. His intellectual interests were protestations, too: he worried that language could be used to dream up problems and solutions in philosophy, and he thought the antidote to dreamy philosophizing lay in tying it to the methods of science.

But science, as Russell himself pointed out, often proceeds without clear inferential rules. To model inference on science, then, we must expose errors in our thinking about scientific investigation and truth-seeking generally. Thus the problem of induction came under his scrutiny; he called it one of the core "problems of philosophy" (in his book of that title), and argued, like Sir Karl Popper, that science does not accrue knowledge by collecting or enumerating facts. In other words, we don't gain scientific knowledge solely by induction. In fact, induction by itself is hopelessly flawed.

Russell offered an obvious and accessible example: observing the sun rise every morning gives us no proof that it will do so again. Our confidence that the sun is coming up tomorrow is no more than a "habit of association," as Hume put it. Induction isn't just incomplete, it positively cannot confirm scientific theories or beliefs by enumerating observations. Our belief that it does gives rise to all manner of distortion. The "gambler's fallacy," for instance, is the wrongheaded belief among gamblers that past frequency of outcome communicates something true about future outcomes. The fallacy can support an expectation of more of the same, or the opposite: time for something new. Streaks when rolling dice create the illusion that the next roll of the dice will be influenced somehow: the good-luck streak is bound to stay good (I'm on a roll), or the bad-luck streak is bound to end (I'm due). Either scenario can happen, of course, but the important takeaway is that the next roll of dice is independent of all prior rolls. The streak continues if the dice randomly fall one way. It ends when they fall another. This is an example of our craving to apply the inductive fallacy even to random events.

Most of the real world is not random, however, which makes rooting out the tendency to see incorrect inductive patterns even harder—the patterns really are "out there," but we don't always know the true ones by observations alone. We see regularities and patterns everywhere. Aside from gambling, this peculiar mental twist helps explain our willingness to generalize from observation. Swans are white. The sun will rise again. The elevator is always waiting for me on the ground floor at 3:30 AM. Of course there are reliable generalizations—we see them everywhere, and it's not delusional to do so—but the problem of induction, as Russell pointed out, is that we have no grounds for inferring knowledge based only on such generalizations. Science must rely on deeper and more powerful inferential strategies. Induction itself is paper-thin.

As an example of the limits of induction, Russell offers us a farmer's well-fed fowl, which is a fabulous inductive thinker. Here is a version of his sad tale:

> This turkey found that, on his first morning at the turkey farm, he was fed at 9 am. However, being a good inductivist, he did not jump to conclusions. He waited until he had collected a large number of observations of the fact that he was fed at 9 am., and he made these observations under a wide variety of circumstances, on Wednesdays and Thursdays, on warm days and cold days, on rainy days and dry days. Each day, he added another observation statement to his list. Finally, his inductivist conscience was satisfied and he carried out an inductive inference to conclude, "I am always fed at 9:00 am." Alas, this conclusion was shown to be false in no uncertain manner when, on Christmas Eve, instead of being fed, he had his throat cut. An inductive inference with true premises has led to a false conclusion.[5]

Russell's turkey exposes the folly of forming "habits of association" without a deeper knowledge of the regularities we observe. But knowledge is often disguised belief—what we think we know can be wrong.

A second and equally damning problem with relying on inductive inference involves lack of knowledge. Much of the world is hidden in mystery—swamped in randomness or chaos, or just too complicated for us to use induction alone with any confidence. Financial markets come to mind. We can try to predict the performance of a stock with all sorts of sophisticated techniques, but as any trader knows, past performance is not indicative of future results. And, if we're honest, much of the world we experience has this frustrating quality. The elevator, we know, settles on the ground floor when no one is using it, and by induction we might expect that it'll be there waiting for us if we're home from work early, because it's off-hours for everyone else. But someone might be moving in, or so-and-so has family visiting from Minnesota, and so on. Rules are made to be broken, and expectations, too.

Our predictions are constantly frustrated because the knowledge we need to augment induction is often lacking or unavailable. I might see a thousand white swans in England and conclude *All swans are white*. That same year, on a trip to Australia, I see a black swan—induction be damned. Much of what we think we know is actually tentative, awaiting further review, and it's overreliance on induction that makes changes seem surprising. In large cities in the western United States like Seattle, drivers typically slow or stop at a yellow light, instead of gunning it to get through. They defer to pedestrians, too, instead of driving around them. I might then be surprised at driving behavior in New York City, or Mumbai. Even when the posted rules are the same, the behavior isn't. If I rely on data and induction from past experience, I might get rear-ended or honked at.

So, why should I rely solely on past events, with all this new information? And what is reasonable and intelligent for me? How should I treat the new information? And what if I see a strip of nails across the road, which I've never experienced, or a line of ducks crossing, or unfamiliar signs? Alas, the answer here isn't more induction, it's less.

INDUCTION WORKS ON GAMES, NOT LIFE

The real world is a dynamic environment, which means it's constantly changing in both predictable and unpredictable ways, and we can't enclose it in a system of rules. Board games, though, *are* enclosed in a system of rules, which helps explain why inductive approaches that learn from experience of gameplay work so well. AlphaGo (or its successor AlphaZero) uses a kind of machine learning known as deep learning to play the difficult game of Go. It plays against itself, using something called deep reinforcement learning, and induces hypotheses about the best moves to make on the board given its position and the opponent's. The approach is fabulously successful on "discreet, observable, two-player games with known rules," as AI scientist Stuart Russell points out.[6] Russell might not have been thinking about Russell's turkey, but he should have been: the real problem with games propping up AI is that they permit hypotheses (generalizations from experience) to be formed according to known rules. Ironically, like classic AI before, the rules don't apply to the real world, which is the entire point of the quest to achieve general intelligence.

Computer scientists relying on inductive methods often dismiss Hume's (or Russell's) problem of induction as irrelevant. As the logic goes, of course there are no guarantees of correctness using induction, but we can get "close enough."

This response misses the point. A method known as "probably approximately correct" governs hypothesis formation for statistical AI, like machine learning, and is known to be effective for weeding out bad or false hypotheses over time. But this method is really an extension of Hume's original argument that induction can supply no guarantees of correctness, as applied to scenarios like games, which have rules to box in statistical inferences. A probably approximately correct solution leaves unchanged the problem of induction in dynamic environments outside a game world or research laboratory.

AI researchers are aware of the problem of induction (either explicitly or implicitly), but it rarely enters into critiques of machine learning (or deep learning) because they are essentially standing the problem on its head. Since induction is bad in dynamic environments, they concede, we apply it in controllable ones. This is like looking for your keys under a lamppost because the light is better there. It's true that human beings have "solved" the problem of induction well enough to use experience effectively in the real world (where else?). But humans solve the problem of inference not with inductive inference in some stronger form, but by combining it somehow with more powerful types of inference that contribute to understanding. Machine learning is only induction (as will be discussed in Chapter 11), and so researchers in the field should be more skeptical than they typically are about its prospects for artificial general intelligence.

REGULARITY AND BRITTLENESS

Induction casts intelligence as the detection of regularity. Statistical AI excels at capturing regularities by analyzing data, which is why visual object recognition tasks like identifying photos of human faces or pets count among its successes. Pixels of faces are distributed and regular in such a way that they can be learned and classified. However, because such systems learn from observations of specific patterns of input, they suffer from problems of brittleness. As Gary Marcus, Ernest Davis, and other researchers have pointed out, even seemingly benign changes, like switching background color from white to blue on object recognition tasks, can degrade performance. Cluttering up photos with other images also results in severe degradation.[7] A few irrelevant letters added to the red area of a stop sign are easily ignored by humans, but when an image altered in this way was presented to one deep learning system, it classified it as a speed limit sign. And there are similar real-world examples, including autonomous navigation systems on self-driving cars that have misclassified a school bus as a snowplow, and a turning truck as an overpass.

Machine learning is inductive because it acquires knowledge from observation of data. The technique known as deep learning is a type of machine learning—a neural network—that has shown much promise in recognizing objects in photos, boosting performance on autonomous vehicles, and playing seemingly difficult games. For example, Google's DeepMind system learned to play a number of classic Atari video games to much fanfare. It was heralded as general intelligence, because the same system was able to master different games using the so-called deep reinforcement learning approach that powered AlphaGo and AlphaZero. But the AI startup Vicarious, for one, soon pointed out that seemingly innocuous changes to the games degraded the ostensibly fabulous performance of the system. In *Breakout,* for instance, a player moves a paddle back and forth on a base line, batting a ball upward into a many-layered brick wall. Each hit destroys one brick (and gets the player closer to "breaking out") but as the ball ricochets back the player must take care not to miss it. Moving the paddle a few pixels closer to the bricks results in severe performance degradation. "DeepMind's entire system falls apart," observe Marcus and Davis in their critique of modern AI. They quote AI pioneer Yoshua Bengio's observation that deep neural networks "tend to learn statistical regularities in the dataset rather than higher-level abstract concepts."[8]

What is often ignored or misunderstood is that these failures are fundamental and can't be patched up with more powerful learning approaches that rely on inductive (data- or observation-based) inference. It is the type of inference here that is the problem, not the specifics of an algorithm. Because many examples are required to boost learning (in the case of Go, the example games run into the millions), the systems are glorified enumerative induction engines, guided by the formation of hypotheses within the constraints of the game features and rules of play. The worlds are closed by rules and they are regular—it's a kind of bell-curve world where the best moves are the most frequent ones leading to wins. This isn't the real world that artificial general intelligence must master, which sits outside human-engineered games and research facilities. The difference means everything.

Thinking in the real world depends on the sensitive detection of abnormality, or exceptions. A busy city street, for example, is full of exceptions. This is one reason we don't have robots strolling around Manhattan (or, for another reason related to exceptions, conversing with human beings). A Manhattan robot would quickly fall over, cause a traffic jam by inadvisably venturing onto the street, bump into people, or worse. Manhattan isn't Atari or Go—and it's not a scaled-up version of it, either. A deep learning "brain" would be (and is) a severe liability in the real world, as is any inductive system standing in for genuine intelligence. If we could instruct Russell's turkey that it was playing the "game" of avoiding becoming dinner, it might learn how to make itself scarce on Christmas Eve. But then it wouldn't be a good inductivist turkey; it would have prior knowledge, supplied by humans.

Statistical AI ends up with a long-tail problem, where common patterns (in the fat head of a distribution curve) are easy, but rare ones (in the long tail) are hard. Unfortunately, some inferences made by humanly-intelligent AI systems will be in the long tail, not in the sweet spot of induction from discoverable regularities in closed-world systems. In fact, by focusing on "easy" successes exploiting regularities, AI research is in danger of collectively moving away from progress toward general intelligence. It's not even that we're making incremental progress, because working on easy problems, in practice, means neglecting the real ones (the keys not near lampposts). Inductive strategies by themselves give false hope.

Getting a misclassified photo on Facebook or a boring movie recommendation on Netflix may not get us into much trouble with reliance on data-driven induction, but driverless cars and other critical technologies certainly can. A growing number of AI scientists understand the issue. Oren Etzioni, head of the Allen Institute for Artificial Intelligence, calls machine learning and big data "high-capacity statistical models."[9] That's impressive computer science, but it's not general intelligence. Intelligent minds bring understanding to data, and can connect dots that lead to an appreciation of failure points and abnormalities. Data and data analysis aren't enough.

THE PROBLEM OF INFERENCE AS TRUST

In an illuminating critique of induction as used for financial forecasting, former stock trader Nassim Nicholas Taleb divides statistical prediction problems into four quadrants, with the variables being, first, whether the decision to be made is simple (binary) or complex, and second, whether the randomness involved is "mediocre" or extreme. Problems in the first quadrant call for simple decisions regarding a thin-tailed probability distribution. Outcomes are relatively easy to predict statistically, and anomalous events have small impact when they happen. Second-quadrant problems are easy to predict but when the unexpected happens it has large consequences. Third-quadrant problems involve complex decisions, but manageable consequences. Then there are the "turkey" problems, in the fourth quadrant. They involve complex decisions coupled with fat-tailed probability distributions, and high-impact consequences. Think stock market crashes. Taleb fingers overconfidence in induction as a key factor in exacerbating the impact of these events. It's not just that our inductive methods don't work, it's that when we rely on them we fail to make use of better approaches, with potentially catastrophic consequences. In effect, we get locked into machine thinking, when analyzing the past is of no help. This is one reason that inductive superintelligence will generate stupid outcomes. As Taleb quips, it is important to know how "not to become a turkey."[10]

There are, of course, other limits to prediction that cannot be neatly summed up by exposing the blind spots in induction. Black swans are rare, after all, as are stock market crashes and major wars (and innovations). We can be forgiven for using induction to help illuminate what are opaque and largely unpredictable possibilities anyway, but not for attempts to replace our understanding with data and statistics alone. In some cases, like chaotic natural systems (say, systems with turbulence) we now know that there are inherent limitations to predictability, using any known type of inferential methods. Induction may suggest that the past will resemble the future, but chaos theory tells us that it won't—or at least that we can't determine how. In some cases, statistical analysis, while incomplete, is all we have.

PROBABLE CAUSE

Turing Prize winner Judea Pearl, a noted computer scientist whose life's work has been to develop effective computational methods for causal reasoning, argues in his 2018 *The Book of Why* that machine learning can never supply real understanding because the analysis of data does not bridge to knowledge of the causal structure of the real world, essential for intelligence. The "ladder of causation," as he calls it, steps up from associating data points (seeing and observing) to intervening in the world (doing), which requires knowledge of causes. Then it moves to counterfactual thinking like imagining, understanding, and asking: What if I had done something different?

AI systems using machine learning methods—and many animals—are at the bottom rung of association. At the first level, association, we are looking for regularities in observations. This is what an owl does when observing how a rat moves and figuring out where the rodent is likely to be a moment later, and it is what a computer Go program does when it studies a database of millions of Go games so that it can figure out which moves are associated with higher percentages of wins.[11]

Pearl here does us a favor by connecting observations and data.[12] He also points out that movement up this ladder involves different types of thinking (more specifically, inference). Associating doesn't "scale" to causal thinking or imaginings. We can recast the problem of scaling from artificial intelligence to artificial general intelligence as precisely the problem of discovering new theories to enable climbing this ladder (or, in the present framework, of moving from induction to other more powerful types of inference).[13]

A COMMON SENSE PRIMER

Your parents, or your partner or a friend, may have accused you of lacking common sense, but take heart: you have much more than any AI system, by far. As Turing well knew, common sense is what enables two people to engage in ordinary conversation. The problem of common sense and in particular language understanding, which requires it, has been a signal concern among AI researchers since the field's inception. And it's finally becoming apparent that hype about machine learning isn't getting us any closer. Researchers are showing more recognition of this, and it couldn't come too soon. Marcus and Davis wonder, if computers are so smart, why they can't read, and they point to "Common Sense, and the Path to Deep Understanding" (a chapter in their book).[14] Stuart Russell begins his list of "Conceptual Breakthroughs to Come" with the as-yet mysterious "language and common sense."[15] Pearl, too, acknowledges language understanding as unsolved (and offers his own "mini-Turing test," which requires understanding of causation).[16]

So, to make progress in AI, we must look past induction. (If you're on the association rung of a metaphorical ladder, look up.) Let's do this next—or at least make a start. On our way to the necessity of abductive inference, we should first get into specifics; in particular, machine learning and its input source, big data.

Chapter 11

...

MACHINE LEARNING AND BIG DATA

Learning is "improving performance based on experience."[1] Machine learning is getting computers to improve their performance based on experience.

This definition of the subfield of AI known as machine learning is widely accepted and not particularly controversial. It has remained essentially unchanged since early work on learning algorithms in AI at the dawn of the field. Carnegie Mellon computer scientist Tom Mitchell, a long-time researcher in machine learning, gave a slightly more detailed definition in his *Machine Learning* in 1997: "A computer program is said to learn from experience *E* with respect to some class of tasks *T* and performance measure *P*, if its performance at tasks in *T*, as measured by *P*, improves with experience *E*."[2] Machine learning, in other words, is computational treatment of induction—acquiring knowledge from experience. Machine learning is just automated induction, so we shouldn't be surprised that troubles with inductive inference spell troubles for machine learning. Fleshing out these unavoidable troubles is the point of this chapter.

There are two main types of learning. When humans first label input to indicate the desired output, it's called supervised learning. Otherwise, the system analyzes patterns in the data as is, and this is called unsupervised learning. There is also a middle ground. Semi-supervised learning starts with an initial seed, or small part of the data, that has been prepared by humans, and then extends the seed to more and more data without supervision.

AI scientists in recent years have largely focused on the specific type of machine learning known as deep learning, which has shown impressive results as a supervised learning approach. In what follows, I will discuss supervised learning in some detail, along with deep learning and its applications. As you might expect, supervised learning is a big tent; I'll explore different types of supervised learning to give an overall flavor of what's at issue for AI.

Classification is a common type of supervised learning. It has been extensively investigated in research labs and in commercial applications. Learned classifiers filter spam, for example. The output is a binary yes or no: either an email is spam or it is not. Typically, the spam classification system is supervised by the user of the email account as he or she marks incoming emails as spam, sending them to the spam or junk folder. In the background, the machine learning system tags the emails that are positive examples of spam. After enough examples of spam are gathered, the system trains itself using them and other incoming emails, creating a feedback loop that converges on the difference between acceptable emails and spam.

The spam filter is one of the earliest examples of machine learning's usefulness on the web. Naïve Bayes algorithms and other simple probabilistic classifiers assign numerical scores to the words in emails indicating spam or not, and the categories of spam and not-spam are supplied by the user. Eventually, the classifier has a hypothesis or model of spam based purely on analyzing words in messages. Future messages get filtered automatically, and the spam goes into its folder. Spam classifiers today use lots of human-supplied knowledge—hints as to what constitutes spam like certain words in the subject line, known "spammy" terms and phrases, and so on. The systems aren't perfect, largely because of the constant cat-and-mouse game between service providers and spammers endlessly trying new and different approaches to fool trained filters.[3]

Spam detection is not a particularly sexy example of supervised learning. Modern deep learning systems also perform classification for tasks like image recognition and visual object recognition. The well-known ImageNet competitions present contestants with a large-scale task in supervised learning, drawing on the millions of images that ImageNet has downloaded from websites like Flickr for use in training and testing the accuracy of deep learning systems. All these images have been labeled by humans (providing their services to the project through Amazon's Mechanical Turk interface) and the terms they apply make up a structured database of English words known as WordNet. A selected subset of words in WordNet represents a category to be learned, using common nouns (like dog, pumpkin, piano, house) and a selection of more obscure items (like Scottish terrier, hussar monkey, flamingo). The contest is to see which of the competing deep learning classifiers is able to label the most images correctly, as they were labeled by the humans. With over a thousand categories being used in ImageNet competitions, the task far exceeds the yes-or-no problem presented to spam detectors (or any other binary classification task, such as simply labeling whether an image is of a human face or not). Competing in this competition means performing a massive classification task using pixel data as input.[4]

Sequence classification is often used in natural language processing applications. Words are treated as having a definite order, a sequence. Document or text classification might use a simple, order-free approach—like a BOW, or bag of words, model—but additional information about the words viewed as an ordered text typically boosts performance in text classification. For instance, words featured in a title and first paragraph often provide strong clues to the meaning or topic of the article. Text classification can exploit these features when auto-tagging articles with topic labels like SCIENCE, BUSINESS, POLITICS, and SPORTS. Text classification is another example of supervised learning, because it begins with humans first tagging articles accurately with topics, providing initial input to the learning system. Like ImageNet's collection of correctly labeled pictures, there are also large corpora, or datasets, that have been created by humans annotating collections of texts, providing metadata about their topics and other features helpful for training supervised learning systems on language-processing tasks.

Supervised machine learning is behind much of the modern web. For example, it enables personalization of news and other content feeds. If a user clicks mostly on political news, then a supervised learning algorithm running in the background (on, say Facebook's servers) will present more and more news stories about politics. More sophisticated approaches classify political news by point of view, offering more conservative or liberal-leaning news stories to a user whose tendency has been identified by the system, and even by sentiment—as when a system classifies opinion text like movie reviews as positive or negative.

In addition to classification, supervised learning approaches are also used for the automatic tagging of individual items in a sequence, rather than of the entire sequence as is the case with image or text classification. This is known as sequential learning. A simple (but maybe boring) example is part-of-speech tagging, where a sequence of words like "the brown cow" is tagged for parts of speech: The / DT brown / ADJ cow / NN, with those tags standing for determiner, adjective, and common noun. Sequential learning does not use linguistic rules to supply programs with knowledge about parts of speech; instead, humans simply tag words in sentences with the correct parts of speech, and provide the human-prepared data as input to the learning algorithm. The part-of-speech tagging problem was solved by machines long ago; supplying tens of thousands of marked-up sentences yields human-level performance on unseen data—that is, any sentences that were not used for training. Another well-explored problem in language processing is named entity recognition, where the supervised learning system predicts typed entities, like mentions of people, places, times, companies, and products in texts. For the sentence "Mr. Smith reported that XYZ Co. sold more than ten thousand widgets in Q1" might be tagged "Mr. Smith / PERSON reported that XYZ Co. / COMPANY sold more than ten thousand / NUMBER widgets / PRODUCT in Q1 / DATE."

Sequential classification can also be used for time-series prediction, where the next item is predicted from previous items. Voice recognition systems like Siri are a kind of time-series prediction, as are popular speech-to-text systems. Time-series prediction has important applications in complex tasks like medical diagnosis, factory planning, and stock prediction, among others.

Supervised learning accounts for nearly all the major successes in machine learning to date, including image or voice recognition, autonomous navigation with self-driving cars, and text classification and personalization strategies online. Unsupervised learning has the virtue of requiring significantly less data preparation, since labels aren't added to training data by humans. But as a direct consequence of this loss of a human "signal," unsupervised systems lag far behind their supervised cousins on real-world tasks. Unsupervised learning is helpful for open-ended tasks like enabling humans to visualize large amounts of data in clusters, which are created by unsupervised learning algorithms. But as most of the ballyhoo about machine learning, and in particular about deep learning, involves supervised learning, I'll focus this discussion mostly on it. Keep in mind, however, that all inductively based limitations of supervised learning approaches apply even more so to unsupervised learning. By focusing on supervised learning, we're looking at a best and most powerful case.

MACHINE LEARNING AS SIMULATION

Machine learning, viewed conceptually and mathematically, is intrinsically a simulation. The designers of a machine learning system examine a data-intensive problem, and if there's some possible machine learning treatment of it they deem it to be "well-defined." They assume that some function can simulate a behavior in the real world or actual system. The actual system is assumed to have a hidden pattern that gives rise to the output observable in the data. The task is not to glean the actual hidden pattern directly—which would require understanding more than the data—but rather to simulate the hidden pattern by analyzing its "footprints" in data. This distinction is important.

Take another language processing task, known as semantic role labeling. Here the designers of a learning algorithm unpack the meaning of sentences in terms of common questions like who, whom, what, and when. The task of the learning algorithm is to take example sentences as input and create the output of a set of labels to answer such questions by identifying the semantic roles expressed in the sentence. For example, the sentence might feature an agent performing an action, an action, a theme (the object involved in the action) and a recipient of the action, and be labeled thus: "John / AGENT threw / ACTION 'the ball' / THEME 'to Lizzy' / RECIPIENT." In all such cases, a machine learning approach involves assuming some actual but unknown behavior, and using a learning approach to simulate it as closely as possible, by learning a function *f*. The result of training the system is the generation of *f* as a model or theory of the behavior in the data. This may model the semantic roles, or entities, or parts of speech, or images of goldfish—it all depends on the learning task. Machine learning is inherently the simulation of a process that is too complicated or unknowable, in the sense that ready-made programming rules aren't available, or that would take too much human effort to get right. Patterns sometimes emerge from data after unsupervised learning reveals them. But the humans identify the pattern after analysis; the algorithm doesn't know to look for it. If it did, that would be supervised learning.

Most of us know about functions from math class, and the classic example is arithmetic: $2+2=4$ is an equation whose operator, the plus sign for addition, is technically a function. Functions return unique answers given their input: thus the addition function returns 4 for $2+2$ (and not ever 5—except in Orwell novels). Early AI scientists assumed many problems in the real world could be solved by supplying rules amounting to functions with known outputs. as with addition. It turned out, however, that most problems that count as interesting to AI researchers have unknown functions (if there are indeed functions involved at all). Hence, we now have machine learning, which seeks to approximate or simulate these unknowns. This "fakeness" of machine learning goes unnoticed when system performance is notably close to a human's, or better. But the simulative nature of machine learning gets exposed quickly when the real world departs from the learned simulation.

This fact is of enormous importance, and gets obscured too often in discussions of machine learning. Here's another fact: the limits of a machine learning system's world are precisely established by the dataset given to it in training. The real world generates datasets all day long, twenty-four hours a day, seven days a week, perpetually. Thus any given dataset is only a very small time slice representing, at best, partial evidence of the behavior of real-world systems. This is one reason why the long tail of unlikely events is so problematic—the system does not have an actual understanding of the real (versus simulated) system. This is enormously important for discussions of deep learning and artificial general intelligence, and it raises a number of troubling considerations about how, when, and to what extent we should trust systems that technically don't understand the phenomenon they analyze (except as expressed in their datasets used in training). We will revisit these themes in later chapters, as they are central to understanding the landscape of the myth.

There are at least two problems with machine learning as a potential path to general intelligence. One, already touched on, is that learning can succeed, at least for a while, without any understanding. A trained system can predict outcomes, seemingly understanding a problem, until an unexpected change or event renders the simulation worthless. In fact, simulations that fail, as they so often do, can be even worse than worthless: think of using machine learning in driving, and having the reliance on automated predictions instill false confidence. This happens everywhere; the messy real world is always changing course. Conversation switches topics. Stocks follow an upward trend, then some exogenous event like a corporate restructuring, an earthquake, or a geopolitical instability sends them downward. Joe may love conservative bloggers until the day his friend Lewis suggests a left-leaning zine, which his personalized news feed has all but screened out and hidden from him. Mary may love horses until Sally, her own horse, dies and move on to pursue a passion for Zen. And so on. Machine learning is really a misnomer, since systems are not learning in the sense that we do, by gaining an increasingly deep and robust appreciation of meaning in the world. They are rather learning bell curves—purely data-driven simulations of whatever we experience directly in the real world.

Common sense goes a long way toward understanding the limitations of machine learning: it tells us life is unpredictable. Thus the truly damning critique of machine learning is that it's backward-looking. Relying on observations from datasets—prior observations, that is—it can uncover patterns and trends we find helpful. But all machine learning is a time-slice of the past; when the future is open-ended and changes are desired, systems must be retrained. Machine learning can only trail behind our flux of experience, simulating (we hope) helpful regularities. It's mind—not machine—that leads the way.

MACHINE LEARNING AS NARROW AI

The simulative nature of machine learning also helps explain why it's perpetually stuck on narrowly defined applications, showing little or no progress toward artificial general intelligence. Well-defined problems in natural language processing such as text classification, part-of-speech tagging, syntactic parsing, and spam recognition, among many others, must be individually analyzed. Systems must be largely redesigned and ported to solve other problems, even when similar. Calling such systems *learners* is ironic, because the meaning of the word *learn* for humans essentially involves escaping narrow performances to gain more general understanding of things in the world. But chess-playing systems don't play the more complex game of Go. Go systems don't even play chess. Even the much-touted Atari system by Google's DeepMind generalizes only across different Atari games, and the system still couldn't learn to play all of them. The only games it played well were those with strict parameters. The most powerful learning systems are much more narrow and brittle than we might suppose. This makes sense, though, because the systems are just simulations. What else should we expect?

The problems with induction noted above stem not from experience per se, but from the attempt to ground knowledge and inference in experience exclusively, which is precisely what machine learning approaches to AI do. We should not be surprised, then, that all the problems of induction bedevil machine learning and data-centric approaches to AI. Data are just observed facts, stored in computers for accessibility. And observed facts, no matter how much we analyze them, don't get us to general understanding or intelligence.

One modern twist on this accepted truth about scientific investigation (and philosophical investigation) is the relatively recent availability of massive amounts of data, which at least initially were thought to empower AI systems with previously unavailable "smarts" and insight. In a sense, this is true, but not in the sense necessary to escape problems of induction. We turn to big data next.

THE END OF BIG DATA

Big data is a notoriously amorphous idea that refers generally to the power of very large datasets to enable analyses and insights essential for businesses and governments (and AI researchers). The actual term first appeared in print in 1997 in a scientific context—in a NASA paper describing challenges to visualizing data using existing computer graphics technology. It didn't catch on, however, until it became popular in the next decade as a catchall term for business and computing. The modern concept of big data appears to have surfaced first in business intelligence discussions, notably in a 2001 Gartner Group report on business intelligence challenges. The report highlighted "three Vs"—volume, velocity, and variety—to describe features of large datasets that would become increasingly important as computational resources continued to get more powerful and cheaper. Yet the report did not actually use the term big data.[5] Nevertheless, the term started appearing everywhere by the end of the 2000s, and by 2014 *Forbes* captured the hype and confusion with an article titled "12 Big Data Definitions: What's Yours?"[6]

It may be hard to precisely define, but big data—orders of magnitude larger data collections—sits at the vanguard of the computational revolution in science and industry. In 2012, the Obama Administration announced a Big Data Research and Development Initiative, intended to "solve some of the Nation's most pressing challenges."[7] And at least one company, the business-analytics firm SAS, quickly invented a new executive title: Vice President of Big Data. Hype, sure. But the excitement about big data was also recognition that more data often spelled more advantage in analyzing problems on increasingly powerful computers.

Yet from the beginning, there was a conceptual confusion about exactly how big data "empowered" insights and intelligence. At first it was thought that big data itself was responsible for better results, but as machine learning approaches took off, researchers started crediting the algorithms. Deep learning and other machine learning and statistical techniques resulted in obvious improvements. But the algorithms' performance was tied to the larger datasets. Regardless, AI was demonstrating performance improvements, and some problems that couldn't be solved at all were suddenly solvable with more data. It was this expansion of insight—in business and in science—that researchers and pundits wanted to capture. As computer scientists Jonathan Stuart Ward and Adam Barker of the University of St. Andrews put it, "big data is intrinsically related to data analytics and the discovery of meaning from data."[8] AI had been struggling to discover meaning from data for decades; suddenly, just by adding more data, such meaning seemed to be revealing itself everywhere.

By 2013, Viktor Mayer-Schönberger and Kenneth Cukier were admitting in their best-selling *Big Data: A Revolution That Will Transform How We Live, Work, and Think* that "there is no rigorous definition of big data," but suggesting anyway that big data is the "ability of society to harness information in novel ways to produce useful insights or goods and services of significant value," and that the arrival of big data means there are now "things one can do at a large scale that cannot be done at a smaller one, to extract new insights or create new forms of value."[9] They point out success stories in the private and public sectors made possible only by increases in dataset size. Take, for example, the startup business Farecast, founded in 2004 by entrepreneur and University of Washington computer science professor Oren Etzioni, which sold to Microsoft in 2008 for over $110 million. Etzioni, who is now head of the Allen Institute for Artificial Intelligence in Seattle, used big data in the form of nearly 200 billion flight-price records to ferret out trends in airfare peaks and valleys as a function of days prior to departure. The performance of Farecast underscored the feeling that big data meant new insights and capabilities emerging from large numbers; progressing from Etzioni's baseline system, which used only twelve thousand price points, the system continually improved its predictions. When it hit billions of flight-price data points, it offered significant customer value in the form of accurate predictions about when to purchase airline tickets.

Big data, once a buzzword, is now the new normal in AI-powered businesses everywhere. Walmart created Walmart Labs to apply big data and data mining techniques to its logistics challenges—buying, stocking, and shipping merchandise efficiently in response to consumer demand. Amazon was using big data before it was a buzzword, tracking and cataloging online purchases, which now are used as data to feed machine learning algorithms offering product recommendations, enhanced search, and other customer features. Big data is an inevitable consequence of Moore's law: as computers become more powerful, statistical techniques like machine learning become better, and new business models emerge—all from data and its analysis. What we now refer to as data science (or, increasingly, AI) is really an old field, given new wings by Moore's law and massive volumes of data, mostly made available by the growth of the web.

Governments and nonprofit organizations quickly joined in, using big data to predict everything from traffic flows to recidivism among parole-eligible prisoners. Mayer-Schönberger and Cukier recount how big data experts from the University of Columbia were hired to construct a predictive model of likely manhole explosions in New York City. (There are over fifty thousand manholes in Manhattan alone). That project was a success, and was offered up as an example of how new insights and capabilities are made possible by increasing the scale of data. Human workers, after all, cannot check tens of thousands of individual manholes each day. Other realms, too, from medical record processing and government actuarial efforts to voting and law enforcement offer examples apparently supporting the claim that new insights and capabilities are made possible by the size and quality of data—big data.

The success of big data in industry and other sectors quickly led to overblown claims about the inferential power of data alone. By 2008, *Wired* editor Chris Anderson's provocation that big data spelled the end of theory in science was a high-water mark for hype.[10] Scientists and other members of the intelligentsia quickly pointed out that theory is necessary, if only because a dataset can't think and interpret itself, but the article stood as a kind of cultural expression for the dizzying success of a data deluge. In truth it was because there was, initially, a hodgepodge of older statistical techniques in use for data science and machine learning in AI that the sought-after insights emerging from big data were mistakenly pinned to the data volume itself. This was a ridiculous proposition from the start; data points are facts and, again, can't become insightful themselves. Although this has become apparent only in the rearview mirror, the early deep learning successes on visual object recognition, in the ImageNet competitions, signaled the beginning of a transfer of zeal from big data to the machine learning methods that benefit from it—in other words, to the newly explosive field of AI.

Thus big data has peaked, and now seems to be receding from popular discussion almost as quickly as it appeared. The focus on deep learning makes sense, because after all, the algorithms rather than just the data are responsible for trouncing human champions at Go, mastering Atari games, driving cars, and the rest. And, anyway, big data has found a new home in modern AI, as data-driven approaches like machine learning all benefit from huge volumes of data for training models and testing them. As one observer put it recently, big data has become "Big Data AI."[11]

So much for big data. But we are left still with the question of inference, and in particular with the question of how data-driven methods like machine learning can overcome the gap between shallow, data-driven simulation and actual knowledge acquired by inferential powers more powerful than induction. The immediate problem is that machine learning is inherently data-driven. I've made this point above; in what follows, I will make it more precisely.

THE EMPIRICAL CONSTRAINT

Data-driven methods generally suffer from what we might call an empirical constraint. To understand this constraint we should put in place one more technical detail of machine learning, known as feature extraction. AI scientists tackling a particular problem typically start by identifying syntactic features, or evidence, in datasets that help learning algorithms home in on the desired output. Feature engineering is essentially a skill, and big money is paid to engineers and specialists with a knack for identifying useful features (and also the talent to tune parameters in the algorithm, another step in successful training). Once identified, features are extracted during training, test, and production phases, purely computationally. The purely computational constraint is the crux. Deep learning systems would perform much better on difficult image recognition tasks if we could simply draw an arrow to the desired object to be identified in a photo cluttered with different objects and backgrounds—using, say, Photoshop software. Alas, the human-supplied feature cannot be added to other photos not prepared this way, so the feature is not syntactically extractable, and is therefore useless. This is the germ of the problem. It means that features useful for machine learning must always be in the data, and no clues can be provided by humans that can't also be exploited by the machine "in the wild" when testing the system or after it is released for use.

Feature extraction is performed in the first, training phase, and then again after a model has been trained, in what's called the production phase. During the training phase, labeled data is provided to the learning algorithm as input. For example, if the objective is to recognize pictures of horses, the input is a photo with a horse in it, and the output is a label: HORSE. The machine learning system ("learner") thus receives labeled or tagged pictures of horses as input-output pairs, and the learning task is to simulate the tagging of images so that only horse images receive the HORSE label. Training is continued until the learning produces a model—which is a statistical bit of code representing the probability of a horse given the input—that meets an accuracy requirement (or doesn't).

At this point, the model produced by the learner is used to automatically label new, previously unseen images. This is the production phase. A feedback loop is often part of production, where mislabeled horse images can be corrected by a human and sent back to the learner to retrain. This can go on indefinitely, although the accuracy improvements will taper off at some point. User interaction on Facebook is an example of a feedback loop: when you click on a piece of content, or tag a photo as a friend, you send data back to Facebook's deep learning–based training system, which perpetually analyzes and modifies your click stream to keep modifying or personalizing future content.

The empirical constraint is a problem for machine learning because all the additional information you might want to supply to the learner can't be used. Unlike image recognition tasks, which rely on pixel data as features, many problems in language understanding incorporate additional mark-up—that is, human-identified features to be extracted by systems when training and using models.

Consider a simple problem in natural language processing, named entity recognition, where some set of semantic tags or labels like PERSON, ORGANIZATION, PRODUCT, LOCATION, and DATE are intended labels or outputs, and the input is free-form text, perhaps from Facebook posts. A company or individual might want to know about all posts that mention a certain company—say, Blue Box, Inc. A keyword search of "Blue Box, Inc." that matches only the words might ignore more informal references in posts like "Blue Box" or "blue box" or even "the box," depending on context. The point of named entity recognition is to use machine learning on large quantities of labeled posts so that these informal phrasings are also, correctly, identified as references to the company. Thus, there is a need for feature extraction: a human labeling all mentions of Blue Box, Inc. in a collection of posts used for training data, and sending it to the system, which generates a model for tagging "Blue Box, Inc." mentions in the production phase.

The Blue Box system relies on words in the posts but also on features—evidence in the posts of mentions of companies. Again, features are necessarily syntactic, because they must be extracted completely automatically in the production phase. This is the key empirical constraint. Features might be orthographic, like checking for capitalization, or lexical, like checking for the occurrence of the words "blue" and "box" in order, and they might include information like "ending with Inc. or Incorporated." A part-of-speech tagger may be run on the training data to tag it with parts of speech like nouns or proper nouns—more syntactically detected features. Other features are no doubt possible. The key again is that all features, while initially identified by humans, are then purely computationally extracted—otherwise the system would not perform automatically during the autonomous production phase.

Here is the problem. Some evidence for mentions of the company Blue Box, Inc. will themselves require inference—say, when pronouns or other references appear in data. This immediately complicates the learning task. If I read a Facebook post and notice that someone is talking about Blue Box and then later refers to it when commenting, say, on "the company's profits," the description "the company's" is not an allowable feature for the named entity recognition system. A co-reference resolution subsystem must handle it, which introduces an error rate—note that co-reference is a much more difficult problem than named entity recognition. Worse, maybe we just happen to know that Bob is talking about Blue Box, Inc. in his discussion of stock performance, but since we can't find this in the data to be analyzed at all, there is no feature to be detected by the system. A person might tell some anecdote about how the founder of some other company, XYZ, Inc. "loved the color blue, and wanted something simple and memorable, and so decided to name the operating system 'Blue Box.'" By context, "Blue Box" is used as the product, not the company, but the named entity recognition system cannot use this contextual information during training. Why? Because it then can't extract it purely by syntax alone, from its input during production.

The empirical constraint is part and parcel of machine learning. It means that only purely syntactical features discoverable in data by automatic methods can be used in training. The truly intelligent system needs features or evidence in a larger sense, not simply from processed data.

Though named entity recognition is a relatively simple task in natural language processing, even here we see the inherent limitations of purely data-driven approaches. A mention of Blue Box in a post about the *product* easily becomes a false positive, and gets labeled as about the company. These examples might be out on the long tail of unlikely occurrences, but they're common enough in ordinary language, and cannot be addressed by machine learning limited by the empirical constraint. All of this is to say that data alone, big data or not, and inductive methods like machine learning have inherent limitations that constitute roadblocks to progress in AI. The problem of induction, it turns out, really is a problem for modern AI. Their window into meaning is tied directly to data, which is a limiting constraint on learning.

THE FREQUENCY ASSUMPTION

In addition to the empirical constraint, machine learning approaches rely on an unfortunate frequency assumption. Like the empirical constraint, this is again a straightforward consequence of—really, a restatement of—the enumerative basis of inductive inference. Ironically, the value of big data for machine learning is actually an exposition of the assumption: more is better. Machine learning systems are sophisticated counting machines. To continue the Blue Box example, we might encode a list of features by checking, say, if the current word or two-word sequence in some Facebook post is in a dictionary of words that includes company names, like IBM, Microsoft, Blue Box, and so on, or checking whether it's followed by Inc. or LLC, or it's an acronym, or the first letter is capitalized, or it's a proper noun.[12]

The frequency assumption comes into play because, in general, the greater the frequency of hits on this feature, the more useful it is for training. In data science, this is necessary; if the features in data are just random, nothing can be learned (recall the earlier discussion of this). But if a pattern does exist, then first, by the empirical constraint, it must be in the data; and second, as a consequence, the only way to determine the strength of association between input and output is by frequency. What else could it be? If every time "Inc." follows a pair of words, the label in the training data is COMPANY, the learner attaches a high probability to "Inc." as a feature of the desired output, company. Patterns that might be undetectable in thousands of examples crystallize in millions. This is the frequency assumption.

Frequency assumptions can be stood on their head, as with so-called anomaly detection, where fraudulent bank transactions or improper logins are detected. These systems, too, rely on the frequency assumption, exploiting what we might call a normality assumption. Normal events make abnormal events more conspicuous. If thousands or millions of examples of proper logins by employees can be grouped or clustered, the odd ones sitting outside the cluster attract attention. They might be illegal or improper attempts, then. Again, machine learning figures out what's normal—and thus what is abnormal—by analyzing frequencies.

The frequency assumption explains "filter bubbles" in personalized content online, as well. Someone who despises right-leaning politics eventually receives only left-leaning opinions and other news content. The deep learning–based system controlling this outcome is actually just training a model that, over time, recognizes the patterns of the news you like. It counts up your clicks and starts giving you more of the same. The same observations apply to recommendations from Netflix, Spotify, Amazon, and other websites offering personalized search and recommendation experiences. This connection between frequency of example (or features in examples) and machine learning is intrinsic, in essentially the same sense that inferring that *All swans are white* gets easier, because you gain more confidence as more and more white swans are observed.

The frequency assumption also explains the difficulty of the long tail problem of abnormal or unexpected examples. Sarcasm, for instance, is particularly opaque to machine learning, in part because it's less frequent than literal meaning. Counting turns out to work well for certain obvious tasks online, but works against more subtle ones. If there are millions of examples of angry citizens tweeting "Trump is an idiot!" then anyone sarcastically tweeting "Trump is an idiot" as a comeback to all the "haters" when he outmaneuvers an opponent will be lumped in as yet another instance of the "Trump-idiot" pattern. The learning algorithm isn't in the knowledge business to start with, so the example is just another sequence of words. Sarcasm isn't a word-based feature, and neither is it as frequent as literal meaning. Machine learning is notoriously obtuse about such language phenomena—much to the chagrin of companies like Google. It would love to detect sarcasm when targeting ads. For example, if "Get me some sunscreen!" is a sarcastic comment by someone posting about a blizzard, a context-sensitive ad placement system should try serving up ads for battery-heated socks, instead.

The frequency assumption gets even more pronounced when input is entire news articles, say, with the text classification task mentioned above. The "funny" and "weird" news prevalent on the web as light reading is a nightmare for machine learning, because the meaning of words is not literal. For instance, a machine learning system might classify stories describing "silly" events that, technically, have criminal or lawbreaking references in them as bona fide examples of crime stories. Why wouldn't it? Odd news is less frequent than its straightforward, information-providing counterpart, so it will be less frequent in training sets—and anyway, detecting why it's odd is another problem with empirical constraints. News is recognized as silly or sarcastic when someone understands what the author means, or intends to communicate. The actual words in the story may, by frequencies in training data, point to well-defined categories like politics, sports, crime, and so on. The story cannot be classified, or understood correctly, unless the constituent bits of syntax—the words—are interpreted in a much broader window of meaning. Absent this non-inductive capability, a machine learning system defaults to frequencies, and misses the point. Here, for example, is an Associated Press story once published by *Yahoo! News*:

> Your Tacos Or Your Life!
>
> Fontana, Calif. A hunger for carnitas nearly led to some carnage after a Fontana man was robbed of a bag of tacos at gunpoint.
>
> Police Sergeant Jeff Decker said the 35-year-old victim had just bought about $20 in tacos from a street-corner stand Sunday night and was bicycling home when the suspect confronted him and said "Give me your tacos."

> Decker said the suspect grabbed the bag of food, punched the victim in the face and began to flee.
>
> When the victim demanded his tacos back, the suspect pointed what appeared to be a handgun at the man and threatened to kill him before running away.

A text classification system would easily identify this as a crime story: *suspect, victim, flee, handgun.* To most human readers, though, it comes across as a comic story—at least we don't see it as a typical example of a crime story. Criminal acts get reported because they are seriously concerning, but an opening phrase like "A hunger for carnitas . . ." signals that AP's intent is to report the story as humorous. Even grade-schoolers will pick up on that intent, but AI systems will happily classify the article as another crime story from Fontana, California. Frequency kills humor. Count up the stories that feature victims, handguns, threats, and fleeing suspects in the news. They're crime. The problem with the frequency assumption for such examples is simply that no known fixes using machine learning are available. The meaning of the stories is lost given the approach, which analyzes words as syntax and counts frequencies of words as evidence for categories. The path is a dead end toward artificial general intelligence, even on relatively simple examples like this one.

Here's another AP story picked up by many newspapers:

> Boy, 11, Bites Pit Bull To Fend Off Attack
>
> Sao Paulo, Brazil - An 11-year old boy is in Brazil's media spotlight after sinking his teeth into the neck of a dog that attacked him.
>
> Local newspapers reported on Thursday that Gabriel Almeida was playing in his uncle's backyard in the city of Belo Horizonte when a pit bull named Tita lunged at him and bit him in the left

> arm. Almeida grabbed the dog by the neck and bit back—biting so hard that he lost a canine tooth.
>
> Almeida tells the O Globo newspaper: "It is better to lose a tooth than one's life."
>
> Stonemasons working nearby chased the dog away before it could attack again.

To be sure, the story has a serious side, but it is certainly not only a story about a pit bull attack. It's not a story about a biting match between a Brazilian boy and a dog, either. Since the boy was not seriously hurt, although he lost one of his own teeth biting the dog, it's clear that the reason for publishing the story was not to report on a Brazilian dog attack but rather to highlight the oddness or silliness of the surprise counterattack. The unlikely content—boy bites dog—is what makes the story news. In a case like this, AI and machine learning don't help at all. They hurt. They miss the point entirely. Ostensible artificial general intelligence systems that used only machine learning would be annoying idiots savants, at best.

Fundamentally, the underlying theory of inference is at the heart of the problem. Induction requires intelligence to arise from data analysis, but intelligence is brought to the analysis of data as a prior and necessary step. We can always hope that advances in feature engineering or algorithm design will lead to a more complete theory of computational inference in the future. But we should be profoundly skeptical. It is precisely the empirical constraint and the frequency assumption that limit the scope and effectiveness of detectable features—which are, after all, in the data to be syntactically analyzed. This is another way of saying what philosophers and scientists of every stripe have learned long ago: Induction is not enough.

MODEL SATURATION

Machine learning and big data suffer from another problem, known as saturation, which bedevils hopes of achieving artificial general intelligence. Saturation occurs when adding more data—more examples—to a learning algorithm (or a statistical technique) adds nothing to the performance of the systems. Training can't go on forever returning higher and higher accuracy on some problem. Eventually, adding more data ceases to boost performance. Successful systems reach an acceptable accuracy prior to saturation; if they don't, then the problem can't be solved using machine learning. A saturated model is final, and won't improve any more by adding more data. It might even get worse in some cases, although the reasons are too technical to be explained here.

Model saturation is rarely discussed, particularly since many recent problems continue to benefit from increases in prepared data. But researchers know that saturation is inevitable, and eventually bounds the performance of machine learning systems. Peter Norvig, Director of Research at Google, let slip in *The Atlantic* back in 2013 his worries about saturation: "We could draw this curve: as we gain more data, how much better does our system get?" he asked. "And the answer is, it's still improving—but we are getting to the point where we get less benefit than we did in the past."[13]

As of this writing, Norvig's cautionary comments are seven years old. The ImageNet competitions probably can't use more data—the best systems are now 98 percent accurate (using the standard test measure of getting a target label in a system's top five predictions). But self-driving cars, once thought to be around the corner, are still in a heavy research phase, and no doubt part of the problem is the training data from labeled video feeds, which is not insufficient in volume but is inadequate to handle long tail problems with atypical driving scenarios that nonetheless must be factored in for safety. The models are saturating, as Norvig predicted. New approaches will no doubt be required. Such considerations are one reason why so-called scaling from initial successes to full-blown ones is naive and simplistic. Systems don't scale indefinitely. Machine learning—deep learning—isn't a silver bullet.

Turing, writing in 1950, hoped that computer systems could be made to learn what they didn't know. Machine learning wasn't then an AI term, though simple neural networks were already known to be possible. But Turing had in mind an expanded notion of learning, more like the human one. Machines couldn't be programmed with all the knowledge they needed; some learning had to occur. He thought it might happen with induction. Some propositions, he mused, "may be 'given by authority,' but others may be produced by the machine itself, e.g., by scientific induction." At his midcentury mark, he had abandoned worries about necessary insights lying outside of formal systems. Or, rather, he hoped they could find a home in the new computing machinery.

Yet scientists themselves don't use "scientific induction" in the sense Turing must have meant. They make guesses, then test them, then make more guesses. Turing never mentioned Peirce's work on logical inference. He had, apparently, no substantive knowledge of abductive inference in Peirce's sense.

We are still in search of his learning machines.

Chapter 12

...

ABDUCTIVE INFERENCE

The Origin of Inference as Guessing

Charles Sanders Peirce was working for the US Coast Survey on a scientific problem of some importance. Peirce swung pendulums. Delicate pendulums. The Coast Survey used them to measure variations in Earth's gravity, part of the science of gravimetrics. A field within geodesy, which in the nineteenth century was still a developing discipline, gravimetrics helps in the study of our planet's shape and size. Precise measurements of Earth's topography were needed for everything from breaking ground on new office buildings to waging war. This was Peirce's job.

Gravimetrics also requires precise measurements of time. As it happened, in the summer of 1879, Peirce found himself on a coastal steamer leaving Boston for New York, in possession of an expensive watch for use in his pendulum work. He had picked it out himself and it cost 350 dollars, a huge sum at the time. The Survey had footed the bill. In the morning, it was gone, along with the rest of his possessions.

Peirce was notorious for losing and misplacing expensive equipment all over the world, wherever the Survey sent him to take gravitational measurements. The theft of the watch fit a pattern that had, by degrees, caused friction between him and the US government. The theft of the watch and Peirce's detective work to recover it were thus of great and personal importance to him. Not surprisingly, he later thought it was all a lesson in inference.

Here is how it went, as he recounted the tale. Peirce recalled that after the ship made its morning arrival in New York's harbor, he left in a cab to attend a conference in the city. Arriving there, he realized he had left behind the watch with its fine gold chain as well as his good overcoat, and rushed back to collect them. Both were gone when he returned to the cabin he had occupied, and they were not being held for him by the captain. It was theft, clearly, and had to have been by one of the ship's stewards.

With the captain's help, Peirce was able to assemble all the stewards on the ship's deck, where they stood in a line as he engaged them with banter and looked into their faces one by one. His hope to detect some clear indication of guilt was disappointed: "Not the least scintilla of light have I got to go on," he admitted to himself. Yet as he started to walk away he thought, "But you simply *must* put your finger on the man. No matter if you have no reason, you must say whom you will think to be the thief."

Peirce looped back to look over the group again—and suddenly "all shadow of doubt had vanished."[1] What had happened? He had been working out the details of his "guesswork" inferences—abduction. Here was a guess. A real-life example—if indeed he was right. Peirce turned to the man he'd fingered as the culprit and, after summoning him out of the line to the stateroom, he offered him a settlement: fifty dollars to return the missing items.

"Now," he said, "that bill is yours, if you will earn it. I do not want to find out who stole my watch, if I can help it; because if I did I should be obliged to send him to Sing Sing [the New York prison], which would cost me more than the fifty dollars; and besides I should be heartily sorry for the poor fool who thought himself so much sharper than honest men."[2]

The accused didn't know much about abductive inference, it turns out, or maybe he thought he was calling Peirce's bluff. "Why," said he, "I would like to earn the fifty dollars mighty well; but you see I really do not know anything about your things. So I can't."[3]

Unable to force a confession, Peirce rushed off to the detective agency Pinkerton's (a "formidable" place). He met the head of the New York office, George Bangs, and said he would like to have the man followed, because he was sure to take the watch to a pawnbroker, who would give about fifty dollars for it. Bangs dismissed Peirce's guesswork, preferring to use the usual rules and known methods for homing in on suspects. He sent a detective to check out the stewards and it was discovered that another one of them had a criminal record including pickpocketing and other petty crimes; this was the more likely thief. Yet the detective's surveillance of that man turned up nothing. Peirce, still convinced that his original suspect would have gone to a pawnbroker, now took Bangs's advice to offer a sizable reward, 150 dollars, for information leading to the recovery of his possessions.

Within a day, the ad he placed had its effect; a pawnbroker came forward with the watch. And once Peirce heard that broker's recollection of the man who had sold it, he had his confirmation—it was a perfect description of the steward he'd accused.

Peirce then got an address (Bangs must have been amused), and showed up unannounced at the man's apartment; two women greeted him at the door and immediately threatened to call for the police. Peirce paid no attention but advanced toward a large wooden trunk he'd spotted. At the bottom of the trunk he found his gold watch chain, complete with his binnacle and compass attached.

Meanwhile, one of the women had disappeared to a neighboring apartment; as Peirce tells it, when two young girls opened that door to his knock, he spied a neatly wrapped bundle on top of a piano and "gently pushed beyond them" to recover his overcoat.

Peirce channeled Dupin, maybe; he swore he guessed. Mr. Bangs shrugged.

Peirce concludes the article in which he relates the tale, "Guessing," with a remark that, at first blush, might seem dubious: "I suppose almost everybody has had similar experiences."[4]

We do guess. Our guesses—inferences—are never certain. But the mystery is: why aren't our hunches, our guesses, no more than random stabs at truth?

At the beginning of "Guessing," Peirce asks how "Galileo and the other masters of science" reached the true theories they did after so few wrong guesses. Scientists, and the rest of us, infer explanations from what we know and observe. We want to subsume these inferences into our stream of observations, into the facts. But so much of what we infer is outside the frame of pure observation. Contextual knowledge pervades almost every inference we make. Peirce's use of Galileo to buttress his story is thus apt: scientific discovery is often attributed to meticulously following known methods, but that's not really true. We hide mystery behind method. Galileo guessed, too, just like Peirce on the steamer. In both cases, subsequent investigation proved that the guess was somehow on track.

Peirce likened guessing to an instinct, a selection out of "at least a billion" possible hypotheses of the one that seems right. Holmes meets Watson and asks him if he's just returned from the war because he sees a tan and a limp. A military doctor fresh from the war in Afghanistan, he figures. Just a guess? No—an inference.

When we seek to understand particular facts—like the theft of a watch—rather than regularities, we are inevitably forced into a kind of conjuring, the selection or invention of a hypothesis that might explain the fact. Induction moves from facts to generalizations that give us (never certain) knowledge of regularity. But abduction moves from the observation of a particular fact to a rule or hypothesis that explains it. Abduction is tied closely to reasoning from events to their causes—in Peirce's example, from the event of the theft to its cause, the thief. Sherlock Holmes called this type of reasoning "nothing more than common sense," and to a large degree he's right. But common sense is itself mysterious, precisely because it doesn't fit into logical frameworks like deduction or induction. Abduction captures the insight that much of our everyday reasoning is a kind of detective work, where we see facts (data) as clues to help us make sense of things. We are extraordinarily good at hypothesizing, which is, to Peirce's mind, not explainable by mechanics but rather by an operation of mind which he calls, for lack of another explanation, instinct. We guess, out of a background of effectively infinite possibilities, which hypotheses seem likely or plausible.

We must account for this in building an intelligence, because it is the starting point for any intelligent thinking at all. Without a prior abductive step, inductions are blind, and deductions are equally useless.

Induction requires abduction as a first step, because we need to bring into observation some framework for making sense of what philosophers call sense-datum—raw experience, uninterpreted. Even in simple induction, where we induce a general statement that *All swans are white* from observations of swans, a minimal conceptual framework or theory guides the acquisition of knowledge. We could induce that all swans have beaks by the same inductive strategy, but the induction would be less powerful, because all birds have beaks, and swans are a small subset of birds. Prior knowledge is used to form hypotheses. Intuition provides mathematicians with interesting problems.

When the developers of DeepMind claimed, in a much-read article in the prestigious journal *Nature,* that it had mastered Go "without human knowledge," they misunderstood the nature of inference, mechanical or otherwise. The article clearly "overstated the case," as Marcus and Davis put it.[5] In fact, DeepMind's scientists engineered into AlphaGo a rich model of the game of Go, and went to the trouble of finding the best algorithms to solve various aspects of the game—all before the system ever played in a real competition. As Marcus and Davis explain, "the system relied heavily on things that human researchers had discovered over the last few decades about how to get machines to play games like Go, most notably Monte Carlo Tree Search . . . random sampling from a tree of different game possibilities, which has nothing intrinsic to do with deep learning. DeepMind also (unlike [the Atari system]) built in rules and some other detailed knowledge about the game. The claim that human knowledge wasn't involved simply wasn't factually accurate."[6] A more succinct way of putting this is that the DeepMind team used human inferences—namely, abductive ones—to design the system to successfully accomplish its task. These inferences were supplied from outside the inductive framework.

SURPRISE!

Peirce understood the origins of abduction as a reaction to surprise:

> The surprising fact, *C*, is observed.
>
> But if *A* were true, *C* would be a matter of course.
>
> Hence, there is reason to suspect that *A* is true.[7]

Surprises are out on the long tail of trouble for induction. And abductive inferences seek explanations of particular facts (*A*), not generalizations or laws, like induction. *C*, too, is a particular—a surprising fact. So abduction isn't a generalization at all.

Inferences from particular observations to particular explanations are part of normal intelligence. If Kate, a barista, usually works at the Starbucks on Thursday but not Friday, a computer with knowledge gleaned from prior experience might not expect her on Friday, but would be confronted with a long tail problem if she's working on

Friday, after all. It might be she is working extra hours, or was called in to cover someone who is sick that day. And she might not work on Thursday, because she's sick, or has been transferred to another store, or quit. These are all particular (surprising) facts that might explain her appearance or otherwise. They are commonsense inferences that don't rely on generalizations or expectations. (Criminal investigations, by the way, always begin with surprising facts. Induction might tell us that young males commit most crimes, but the investigator still needs to know who in particular is responsible for this one—and the culprit need not be young or male, or even human, as we saw in the Rue Morgue.)

Peirce understood abduction as a weak form of inference, in the sense that it was conjectural—an abduction at time t might be proven wrong at time $t+1$. Much inference in the real world is defeasible, that is, proven wrong or incomplete by subsequent observation or learning (say, by reading a book).

Conjectural inference is a feature, not a bug, of intelligent systems. Rosie the Robot might believe that Kate has quit Starbucks because a coworker has provided this information, but when Kate shows up for work ten minutes later, and the coworker is smiling, Rosie the Robot should retract its inference. We scarcely notice how quickly we conjecture plausible reasons for what we see (or read about), and also how quickly we drop or update such conjectures. The everyday world is a constant stream of seemingly surprising facts against a backdrop of expectations. Much of the world, like a traffic light, isn't a constant surprise—but then, traffic lights do break, too.

The meaning of an observation itself undergoes a conceptual change with abduction, as well. Whereas induction treats observation as facts (data) that can be analyzed, abduction views an observed fact as a sign that points to a feature of the world. Signs can be thought of as clues, because they are understood from the beginning as embedded in a web of possibility that may help point to or shed light on a particular problem or question important to the observer. In rich cultural contexts like crime-solving, clues are necessary because there are too many facts to analyze, and only a few are relevant. Indeed, the basic problem with using known methods in detective work is that difficult or seemingly insoluble crimes don't fit regularities, and the accumulation of facts doesn't point anywhere. Smart detectives look for clues.

So do hunters. Not only are hunters astute observers—they observe certain kinds of things. Hoof marks, droppings, tufts of hair, broken branches, and scents are all clues to the location of prey. Like detectives at a crime scene, hunters are engaged in a deliberate search for evidence of recent action; they observe results from the past.

Perhaps counterintuitively, clues are not considered unique. A hunter who comes across an unknown scent will assume that it points to something perhaps interesting, but not entirely unique, because if it were utterly unique the scent could not function as a clue. A hunter who reasons that an unknown scent might be from a previously undiscovered species or an extraterrestrial will not advance his or her interests in finding prey.

Thus, the hunter is interested in a conjecture that fits his or her specific purposes. Perhaps the smell is unique because it was produced by an animal during mating season. Hence the scent is a surprising fact that might be accounted for by conjecturing its source from changes in the animal because it's mating season. A familiar animal in mating season exudes different smells. Keep in mind that the hunter has no prior experience of this phenomenon (hence the abductive inference), but reasons within a framework that excludes logical possibilities that don't advance the objective. Such conjectures are, surprisingly, likely to be true—which is why Peirce puzzled over the "guessing instinct" at the core of so much intelligent thought.

Just as hunting is a prime example, so is seeing. Even judging that a particular object is an azalea, as Peirce points out, involves physical perception in a deep network of prior knowledge and expectation.

Recall his words: "Looking out of my window this lovely spring morning I see an azalea in full bloom. No, no! I do not see that; though that is the only way I can describe what I see. That is a proposition, a sentence, a fact; but what I perceive is not proposition, sentence, fact, but only an image which I make intelligible in part by means of a statement of fact. This statement is abstract; but what I see is concrete."[8]

Peirce's insistence that abductive inference undergirds even seemingly trivial visual perceptual abilities might seem to be contradicted by recent successes using convolutional neural networks (deep learning) on visual object recognition tasks, as with the runaway successes on ImageNet competitions. Yet such seeming successes actually prove Peirce's point, as the research community has (to its credit) quickly pointed out the brittleness of these systems in a growing literature that touches on not only central questions of inference but also concerns about trust and reliability, as well as potential for misuse. As computer scientist Melanie Mitchell points out, even the winning deep learning systems are ridiculously easy to fool.[9]

AlexNet, for instance, the system that blew away the field in the 2012 competition, can be tricked into concluding, with high confidence, that images of a school bus, a praying mantis, a temple, and a shih tzu are ostriches.[10] Researchers call these adversarial examples, and they are accomplished by strategically changing a few pixels in the images—so few that the changes are not at all noticeable to the human eye. The images still look exactly like the originals to humans.

So-called adversarial attacks are not unique to AlexNet, either. Deep learning systems showing impressive performance on image recognition in fact do not understand what they are perceiving. It is therefore easy to expose the brittleness of the approach. Other experiments have drastically degraded performance by simply including background objects, easily ignored by humans, but problematic for deep learning systems. In other experiments, images that look like salt-and-pepper static on TVs—random assemblages of black and white pixels—fool deep learning systems, which might classify them as pictures of armadillos, cheetahs, or centipedes. As modern AI progresses, these obvious shortcomings actually prove the depth of knowledge and context that enables even visual perception. Peirce was right, in other words, about seeing an azalea on a fine spring morning, or about seeing anything else: "I perform an abduction when I [do so much] as express in a sentence anything I see. The truth is that the whole fabric of our knowledge is one matted felt of pure hypothesis confirmed and refined by induction. Not the smallest advance can be made in knowledge beyond the stage of vacant staring, without making an abduction at every step."[11]

The origin of intelligence, then, is conjectural or abductive, and of paramount importance is having a powerful conceptual framework within which to view facts or data. Once an intelligent agent (person or machine) generates a conjecture, Peirce explains, downstream inference like deduction and induction make clear the implications of the conjecture (deduction) and provide a means of testing it against experience (induction). The different logics fit together: "Deduction proves that something must be; Induction shows that something actually is operative; Abduction merely suggests that something may be."[12] Yet it's the *may* be—the abduction—that sparks thinking in real-world environments.

The defeasible nature of abduction helps explain its centrally important role in natural language understanding, not just in hunting or detective work. Our understanding of what's being said in ordinary language is constantly subject to update and revision. Consider this snippet of English: *Raymond saw a puppy in the window. He wanted it.* The pronoun *it* probably refers to *puppy* (linguists call this an example of pronominal anaphora—or reaching back). The two sentences are in an "out of the blue" context, and we don't have any more information about Raymond, but common sense reminds us that people typically desire puppies rather than windows, and that we often look through windows to objects that might hold some interest for us. But the inference that *it* refers to *puppy* is not certain. Change the example by adding context, and that inference is incorrect: *Raymond broke his window. He went out shopping for a new one. He told himself he would know when he found the right one. A beautiful storm window was on sale. Raymond saw a puppy in the window. He wanted it. It was the right window for him.*

The example may be contrived, but there is nothing wrong about it. Raymond might be the sort of fellow who puzzles over seemingly mundane purchases. He might be disposed to take the appearance of a puppy as a sign he should get a particular window, maybe out of some superstition. The point is that if we view the pronoun *it* as a sign, it can point or refer back to different nouns as the context changes. All existing strategies in AI to date have failed to adequately account for such examples.

Researchers call deduction "monotonic inference" because conclusions are permanent—once an AI system deduces a conclusion, the conclusion is automatically added to the system's store of knowledge. Yet language understanding is non-monotonic (requiring defeasible inference). New information from successive sentences can force changes in initial interpretations: *It was at that very moment that Raymond knew this was the window for him.* To get the gist of a narrative we have to understand how each new sentence affects interpretation of prior ones. This is built into abduction, which is, after all, conjectural and subject to revision from the get-go.

In classical AI (AI before the web), researchers tried different ways of extending inference to make it defeasible. By far the most common approach involved extending deduction. Work on so-called nonmonotonic reasoning peaked in the 1980s and 1990s, but has since been largely abandoned, in large part because the extensions of deduction to make it more flexible for language understanding work only on "toy" examples that aren't useful in the real world. A classic example is reasoning thus: "If *x* is a bird, then *x* can fly. *X* is, in fact, a bird. So it can fly. Wait! It is a penguin. Penguins can't fly. Therefore, *x* can't fly (after all)." There are defeasible reasoning systems that permit reasoning like this, but they are notoriously intractable in the general case (that is, computationally infeasible), and they have never scaled to handle complicated inferences required for interpreting the ordinary language found in news articles. Non-monotonic inference systems work only on contrived scenarios in the laboratory.

Even if such systems could be scaled, though, the core problem with deduction is its truth-preserving constraint—everything must be certain. We might modify or reject inferences later, but we are wasting time using deduction in the first place.

Many of the same researchers who worked on extending deduction in AI to make it defeasible also worked out deductive-based approaches to abduction in the 1980s and 1990s, notably with abductive logic programing (ALP). Without delving into the technical details, an ALP inference is an entailment (a truth-preserving deductive inference) from a logical theory T (the knowledge base) to the truth of a conditional $E \rightarrow Q$, where E is an explanation of Q, the observation. This is a fancy way of eliminating the conjectural nature of abductions, in effect. No inferential power is gained, which explains why work on ALP, like non-monotonic reasoning strategies, has languished and largely been abandoned. (The problem with ALP aptly demonstrates the core problem with commonsense inference generally, and we'll return to this in an upcoming section.)

If we try to preserve Peirce's conception of abduction as a conjecture to a plausible hypothesis, we end up in inferential "trouble." In particular, we end up with a reasoning mistake. In studies of logic, it's called a fallacy, and we turn to it next.

FALLACIES AND HYPOTHESES

Peirce symbolized abduction as "broken" deduction. It's easiest to see this by considering again the rule *modus ponens* for straightforward deduction:

$A \rightarrow B$ (Knowledge)

A (Observation)

———

B (Conclusion)

And the truth table:

A	→ B	Conclusion
T	T	T
T	F	F
F	T	T
F	F	T

The second row is the issue: *A* is true but *B* is false, which makes the inference from *A* → *B* false. For instance, if *A* represents *It is raining,* and *B* represents *The streets are wet,* then the conditional expression says that it's always true that if it's raining, then the streets are wet. But if A represents *It is raining,* and B represents *The streets are dry,* then the material conditional *A* → *B* (if *A*, then *B*) is false. The second row of the truth table tells us this.

Notice the third row: if it's not raining, but the streets are in fact wet, well, the streets are still wet, so the rule is still truth-preserving. But at any time that it's raining and the streets are not wet, the rule is wrong, so the outcome of applying it will be false. This is standard logic, called "propositional" because the variables stand for full statements, or propositions.

Propositional logic was developed a very long time ago, and it has been proven not to have bugs. It's complete, meaning that anything true in the logic can be proven (using its rules), and anything provable is also true. It's also consistent, because you can't prove a contradiction. If A is true, "not-A" must be false, and the system can't prove the latter if it proves the former. Nothing can really go wrong in the simplified world of propositional logic. You can derive the truth of any propositions that have been expressed in it, and you can never derive nonsense—believing, say, that it's both raining and not-raining. The system is consistent.

Now consider the following fallacy, not allowed in propositional logic or deduction generally:

Affirming the Consequent

$A \rightarrow B$

B

———

A

Logicians call this argument form an example of "affirming the consequent," because B is the consequent of the rule (A is called the antecedent), and B is given as the second premise, the case or fact observed. Clearly, though, when we use the consequent as the case, we err. We make a mistake in reasoning. This follows because A might be false, so we can't conclude that it's true in all situations, as required by deduction. Since deduction is truth-preserving, the inference is invalid. It's a fallacy.

It's easier to see this by assigning actual English-language sentences (propositions) to A and B:

If it's raining, the streets are wet.

The streets are wet.

Therefore, it's raining.

The argument is invalid because, even accepting the truth of the premises, its conclusion is not necessarily true: the street might be wet for other reasons (for example, a fire hydrant might have burst). Affirming the consequent is bad deduction because it's a guess. Viewing the logical form of abduction as a variant of bad deduction helps explain why it has been ignored, historically, in studies of reasoning, and also why it has resisted mechanical methods like those found in AI. How do we incorporate bad rules?

Indeed, Peirce's own formulation of the types of inference makes it clear that we cannot translate abduction into a kind of deduction, for exactly the reason just given. He used syllogisms, expressed in English language statements:

DEDUCTION

All the beans from this bag are white.

These beans are from this bag.

Therefore, these beans are white.

INDUCTION

These beans are from this bag.

These beans are white.

Therefore, all the beans from this bag are white.

ABDUCTION

All the beans from this bag are white.

These beans are white.

Therefore, these beans are from this bag.

Converting this into propositional logic, we have:

Deduction

$A \rightarrow B$

A

———

B

Induction

A

B

———

$A \rightarrow B$

Abduction

$A \rightarrow B$

B

———

A

In other words, abduction by its very nature cannot be an extended form of deduction, because its logical form (essence) is an egregious deductive fallacy. It breaks the truth-preserving nature of deductive inference, which makes sense given that, as Peirce argued, it begins with a conjecture or guess, which by definition might be wrong.

In fact, all three inference types in this framework are distinct: one type cannot be converted into another, which implies that if intelligent inference requires abduction, we cannot get there through deduction or induction. This observation is of paramount importance to work on AI. If deduction is inadequate, and induction is inadequate, then we must have a theory of abduction. Since we don't (yet), we can already conclude that we are not on a path to artificial general intelligence.[13]

I pointed out previously that the conjunction of the frequency assumption and the empirical constraint eliminate induction as a complete strategy for artificial general intelligence. Reliance on frequencies in data gives us Russell's turkey, whose confidence that the farmer cares about it actually increases as observations of care stack up—the day before the Christmas feast is, inductively, the highest probability that its beliefs are correct (since it has the most inductive support). And the empirical constraint is a hard boundary on the knowledge or theory we can give to the turkey. If we tell it to "watch how the farmer treats you," it will get happier and happier until it's dead. But since the observation "I'm on the chopping block" isn't in the data in all prior times until (t = Christmas Day), it can't be supplied to the turkey's theory or model by pure induction.

In machine learning, this means that the only knowledge we can supply to a system is what can be recovered in the data purely syntactically. This has been seen as a virtue, as in the case of DeepMind's Atari system, but it implies the same type of blind spot that plagues Russell's turkey—what it can't observe in the data, it doesn't know. This results in failed predictions, like image-recognition foibles, and it also accounts for the peculiar brittleness of modern systems, where seemingly trivial changes to pixels degrade performance on games and other tasks.

We can patch up purely data-driven, inductive systems by including more data—to some extent. But exceptions, atypical observations, and all sorts of surprises are part and parcel of the real world. The strategy of exposing supervised learning systems to foreseeable exceptions, as is done with ongoing work on driverless cars, is a Sisyphean undertaking, because exceptions by their very nature cannot be completely foreseen. A new, essentially abductive, approach is required. In the meantime, we're stuck with only the observable regularities, the "fat head" of what is automatable.

Judea Pearl has made this point nicely with his metaphor of the "ladder of causation." He calls machine learning and statistics an exercise in "fitting data to a curve" (and technically, it is), which fits into his first rung, of association. We can ask questions about correlation in the first rung, like "What does a survey tell me about the election results?" And we can use correlations between moves in a game and winning outcomes to design modern game-playing systems like AlphaGo. But we cannot extract causal information about the world from associations in data, so explanations involving *why* or *how* questions can't even be formulated, let alone answered. Causal knowledge forms part of our commonsense understanding of the world, and explains why, for instance, we can see data as effects or clues to prior causes that contribute to our understanding.

Importantly, at the top rung of Pearl's ladder are counterfactuals, where we ask *what if* questions whose answers don't exist at all in any dataset (by definition, because we are asking about what hasn't happened). We imagine worlds. In such counterfactuals, part and parcel of human intelligence, data is useless to help us determine whether, say, Kennedy would still be alive if Oswald had never lived, or whether the turkey would be safe if the farmer had received a gift turkey, or if the streets would still be wet if the fire hydrant hadn't been dislodged by a swerving bus. Imagination involves inferences that don't exist in a dataset. And imagining requires conjecture, if anything does. Abduction is inference that sits at the center of all intelligence.

Though researchers of late seem to have forgotten, for most of AI's checkered history the problem of acquiring and using commonsense knowledge about the world has been its core challenge. Common sense requires a rich understanding of the real world, which decomposes broadly into two parts: first, AI systems must somehow acquire everyday knowledge (and lots of it); and second, they must possess some inferential capability to make use of it.

Hector Levesque, an AI scientist at the University of Toronto, makes a good case in his 2018 *Common Sense, the Turing Test, and the Quest for Real AI* that, absent a rich theory of knowledge, our attempts at achieving artificial general intelligence ("real" AI) are doomed.[14] His analysis of the necessity of commonsense knowledge is spot on, but he apparently falls into the traditional trap of understanding inference as an (as yet unknown) extension of deductive reasoning.

Levesque in large part is attempting to resuscitate a once-prominent field in AI known as knowledge representation and reasoning (KR&R), which addressed head-on problems of knowledge and inference in intelligent systems. The reasoning part of KR&R involves considerations of inference, and the problems we've just reviewed have been more or less discovered in the field of KR&R, but unfortunately left without solutions. In an earlier paper called "On Our Best Behavior," in 2013, Levesque pointed out that extensions (and fixes) to deduction falter because they are *intractable,* a computer science term meaning that solutions can't be computed in real time (if at all): "Even the most basic child-level knowledge seems to call upon a wide range of logical constructs. Cause and effect and non-effect, counterfactuals, generalized quantifiers, uncertainty, other agents' beliefs, desires and intentions, etc. And yet, symbolic reasoning over these constructs seems to be much too demanding computationally."[15]

Intractability is one clue that the approach itself is wrong. A deeper clue is simply that deduction cannot be fitted into the logic of abduction. This irreducibility implies that the problem is fundamental, regardless of issues of computational expense.

Both aspects of KR&R—that is, both representation and reasoning—are necessary today. In particular, representing commonsense knowledge in a machine has proven difficult, to put it mildly.

Even after decades of work, no one has come even close to getting enough basic knowledge into a machine to empower it in real-world scenarios, like understanding ordinary language, or navigating around the house or on a busy street.[16] Knowledge and reasoning are obviously interlinked, because we can't infer what we don't know, and we can't make use of knowledge that we have without a suitable inference capability. I call these problems the "bottomless bucket of knowledge" and the "magical inference engine."

Ironically, old efforts on Abductive Logical Programming aptly demonstrate the key stumbling blocks. Suppose (again) that T is some knowledge base for an AI system S, which makes a (surprising) observation Q such that an explanation (or cause) E is in T, and T entails $E \rightarrow Q$, where entails is a strong truth-preserving inference (technically, entailment is stronger than material implication "$\rightarrow$" since every statement in T must also make $E \rightarrow Q$ true, and vice versa). T is thus S's "theory of everything." Question 1: How do we get all necessary knowledge into T? Question 2: Since $E \rightarrow Q$ is just the material conditional we saw above, how does the truth of $E \rightarrow Q$ constitute an abductive inference to a plausible (testable) conjecture, of, in this case, E for observation Q? In other words: How do we get a theory T, and, how do we use it abductively? All that's required to explain these problems is wet pavement.

The US Department of Defense, through its high-tech research arm, the Defense Advanced Research Projects Agency, once invested huge sums of money into the construction of large commonsense knowledge bases. Experts trained in logic and computation (full disclosure—I was one of these experts) spoon-fed computational systems with ordinary statements like *Living humans have heads,* and *Sprinklers shoot out water,* and *Water makes things wet,* and so on. Ostensibly, AI systems armed with gobs of these commonsense facts could draw on them to infer things about the world. Such systems wouldn't need abductive inference, it was hoped, because the systems crammed with knowledge could use simpler and better understood approaches to inference. Since *When the sprinklers are on, the area around them gets wet* is as true as similar statements about rain, the systems' inference engines could stick with well-known deduction, after all. Sensors (or text input) could inform AI systems about an approaching tidal wave, and deduction could infer that the streets got wet.

A number of major challenges to this approach soon surfaced. The first problem was obvious, or should have been: most of what we know is implicit. We bring our knowledge into consciousness, making it explicit, only when circumstances require it, like when we are surprised or have to think through something deliberately.

This brings us to the second problem, the "tip of the iceberg." All our implicit knowledge might be necessary for some inference or other, but the total amount is vast. The knowledge base of an ordinary person is unbelievably large, and inputting and representing it in a computer is a gargantuan task.

Spoon-feeding a computer with common sense turned out to be a lifelong philosophical project, ferreting out commonsense knowledge like *pouring a liquid into a glass container with no cracks and only one opening will fill it up*. Or that *living humans have heads*, or that *a road is a pathway with a hard surface intended for vehicle travel*. Researchers were assuming computers would "get it" eventually, but eventually, the project seemed unending.

Think of an AI system built just to answer intelligently about the admittedly simple and boring topic of wetness. It would itself require a massive knowledge base. Any realistic, say, conversation-ready machine for only this topic would need concepts for firefighting airplanes carrying water (and not fuel, though that would make the streets wet, too), Super Soakers, kids playing, and on and on. Consider a simple question put to the system akin to a Turing test query: "It was a hot day without a cloud in the sky. The fire department was called to shut off a malfunctioning fire hydrant. Main Street was wet, and the gutters had become clogged with debris from rushing water. *Why were the streets wet?*" If the AI system's knowledge base lacks fire departments and hydrants, it has no hope of answering the question. And consider tacking on this additional sentence to the question: "But that wasn't what made the streets wet. The hydrant was spraying onto the front windows of the deli. A massive thunderstorm hit the area right before the fire department showed up. Everything was wet!" How do all those concepts in the knowledge base help? How does deduction?

COMPUTATIONAL KNOWLEDGE AS A BOTTOMLESS BUCKET

This is the "bottomless bucket" problem: filling up a computational knowledge base with statements expressed as propositions (in a logic) is an endless task. We can't solve even simple commonsense problems, like reasoning about happenings on a city block or a neighborhood, without effectively codifying huge volumes of seemingly irrelevant knowledge.

There's also a representation problem. Knowledge bases, like relational databases, must be organized and structured, so that the bits of knowledge relevant for talking about wet streets, or Kate's absence from the coffee shop, or half-eaten cans of tuna, or what have you, are available for inference. It won't do for an AI system to start computing the position of Mars in the night sky, when asked if a spoon was used to eat tuna. This is the problem of relevance, and it has to be cunningly thwarted with strategies for representing knowledge to make it accessible and available. Researchers typically try to "pre-solve" relevance issues by grouping related knowledge together so that the rules (like *modus ponens*, and many others) cover what's necessary.

Typically, a knowledge base is organized into hierarchies using "is-a" links, indicating that something is an instance of something else, like saying a laptop is a PC, which is a type of computer, which is a digital technology, and so on. In large knowledge bases, hierarchies form the backbone of encyclopedic clusters of topics about the things we frequently encounter in the world. In addition to all the "is-a" links, other predicates (that is, links between concepts) expressing other important relationships can be introduced, like part-whole (meronymic) relations between concepts.[17] Knowledge representation languages have evolved since the early years of AI to facilitate knowledge base development, using simplified vocabularies tailored for expressing hierarchical relationships.

In the 1990s, for instance, working groups on the web developed the resource description framework (RDF) for writing "triples" (subject-predicate-object phrases) in which the predicate can be "is-a," "part-of," or anything else deemed useful for building the knowledge base. RDF helped knowledge bases become, in essence, computational encyclopedias (as the late AI researcher John Haugeland once put it) with larger projects' knowledge bases having thousands of triples. AI researchers hoped that the ease of use would encourage even non-experts to make triples—a dream articulated by Tim Berners-Lee, the creator of HTML. Berners-Lee called it the Semantic Web, because with web pages converted into machine-readable RDF statements, computers would know what everything meant. The web would be intelligently readable by computers. AI researchers touted knowledge bases as the end of brittle systems using only statistics—because, after all, statistics aren't sufficient for understanding. The Semantic Web and other knowledge base–centered projects in AI could finally "know" about the world, and do more than just number-crunch. An enormous amount of money and effort was wasted on this dream, which never really worked. The bottomless bucket problem is still with us, because the task is still bottomless.

But since knowledge is, typically, organized by subject matter so it can be accessed quickly by an AI system (which can't, after all, take all day pondering why the streets might be wet when posed a question), engineers found themselves packing more and more "odd" information into a given subject area to cover everything that might come up. To use our example again, firefighters, hydrants, Super Soakers, firefighting planes, rain, snow, floods, and so on all had to, somehow, be connected together for the system to have a prayer of answering basic questions about wet streets. But this strategy quickly stops making sense; it frustrates the original intention to organize the knowledge naturally, by topic. While fire hydrants might be part of a description of, say, a typical city block along with its streets, large firefighting planes, or descriptions of children's toys that shoot water, probably won't. In other words, the very attempt to organize knowledge so it's computationally accessible for real-time reasoning inevitably leaves out items that will be required given some scenario or other. There doesn't appear to be any recipe for constructing a knowledge base for intelligence—at least, not one we know about, thought out in advance, to be filled in with logical languages like RDF or anything else.

The ultimate proof of the insolubility of the bottomless bucket problem reared its head only gradually, after the projects were hopelessly underway. Researchers had committed to massive knowledge base development projects, and thousands of person-hours had been spent filling out actionable encyclopedias to tame the commonsense problem of AI. As more "knowledge" was added, though, systems had more and more chances to get everything wrong. A system that knows only about rain and wetness might get a simple question right, but introducing dozens of different possible scenarios meant that, in general, the system would inherit more and more ways to get it all wrong. *The fire department was called out, because the downpour had flooded several parks,* is actually about rain, which caused the flood. But the reference to the fire department might flummox a system that "knew" about fire hydrants breaking and potentially wetting streets. Thus, unlike human thinking where in general the more we know, the more power we have to make useful inferences, large knowledge-base projects are always in danger of generating nonsense inferences with all their additional "knowledge." Clearly, not only the statements, facts, and rules constituting common sense are needed; how commonsense inference itself works is also important.

Looking in the rearview mirror, early efforts to give AI systems "commonsense knowledge" were actually two projects, disguised as one. Knowledge is an obvious requirement, but inference is, too. What we know is one piece of it; how we use what we know to update our beliefs is quite another. But strategies for computational inference are limited. As Peirce pointed out, we have three, and to date only two have been reduced to computation. Relevance problems with pure deduction surfaced, and so researchers tried out different ad hoc schemes.

In the 1970s, for instance, Roger Schank at Yale developed a "scripts" approach to common sense.[18] Schank argued that ordinary thoughts and actions follow an implicit script—we follow a story line. A paradigm case here is ordering food at a restaurant. We walk in, get seated (or seat ourselves), look at the menu, and order when waitstaff appears, after an initial greeting. All of this occurs in sequence and can be to some extent anticipated and planned. Schank assumed that some activities in the real world could be scripted in this way, and he developed some systems that used programmed scripts to interact with people on tasks like ordering food. In the real world, though, we keep veering off of scripts. We need common sense to order food, or to make a reservation whenever something unexpected happens, like waiting too long for service. Perhaps there's a sign at the front entrance, instructing customers to use the side door.

Schank's work on scripts was abandoned as a serious attempt to solve AI's commonsense knowledge problem, but it is still instructive because it shows clearly how knowledge and inference are separate. We might suppose that all knowledge relevant to ordering food can be given to an AI system—all the concepts and predicates describing a restaurant and ordering food are in the Restaurant Knowledge Base. But disruptions, unexpected events, or just plain conversation with the waiter will quickly confuse a scripts-based system. Invariably, it will need to know a lot more about the world than what's in the Restaurant Knowledge Base.

It's tempting to try to supply all the extra-restaurant stuff in a "meta" knowledge base full of knowledge about all sorts of commonsense things: the weather, the sports game on TV, and so on. But as there won't be direct links to all of this (because of the problem mentioned earlier, that we can't "pre-solve" all the knowledge that might become relevant), the system must somehow jump outside its script of subject-related knowledge and plans. But to do that, it must have some inference mechanism that knows what's going on—where to look, and for what. And to do that, we must solve a problem other than knowledge per se. We need a non-deductive (and non-inductive) flexible, commonsense inference mechanism. This brings us back to inference.

MAGICAL INFERENCE ENGINES: THE SELECTION PROBLEM FOR AI

If we have a rule, say $A \rightarrow B$, and we observe a (surprising) fact B, we might hypothesize that A is true, because it would explain B (since $A \rightarrow B$ is known). Thus if *If it's raining, the streets are wet* is a known true statement, and we see that the streets are wet, knowing nothing else we might suppose it's been raining. Indeed, in an out-of-the-blue context, it's probably most likely.

But a system which also observes C, say "It's a cloudless sky," should ignore the rule $A \rightarrow B$. Why? Well, because it's not relevant anymore. In fact, recalling that A is a variable, it can stand for lots of different statements: *It's raining*, or *The kids are using the Super Soakers*, and so on. In fact, we can extend A to include a set of statements, like this: A = {Rain, FireHydrant, Sprinklers, SuperSoakers, Tsunami, . . .}. Now the task is to select from among the members of the set A, in the antecedent of the rule $A \rightarrow B$, the true statement that is also most relevant to the observation B. Understanding relevance is unavoidable; knowing B tells us nothing about which member of the set in A should be used in $A \rightarrow B$. We must pick what we think is the most relevant member of the set.

Thus the deductive inference from observation A to conclusion B is certain but too easy. We want to observe the effect, and infer the cause: the wet streets, and why, or how. This is all part of normal intelligence, what Marcus and Davis call having a meaningful picture of the world, in which knowledge about what causes what is critical. Acquiring and using this knowledge is complicated, because most real-world events admit of many possible causes. The selection problem is finding the operative or best or plausible cause, given all the possibilities real or imagined. So the core problem of automating abductive inference can be recast as this problem of selection, which helps expose the difficulty inherent in the required inference, but in the end it's really the same problem. To abduce we must solve the selection problem among competing causes or factors, and to solve this problem, we must somehow grasp what is relevant in some situation or other. The problem is that no one has a clue how to do this. Our actual inferences are often guesses, considered relevant or plausible—not deductions or inductions. That's why, from the standpoint of AI, they seem magical.

Peirce offered what he called a "vague explanation" of the unreasonable accuracy of our guessing: "There can, I think, be no reasonable doubt that man's mind, having been developed under the influence of the laws of nature, for that reason naturally thinks somewhat after nature's pattern."[19] He estimated, perhaps conservatively, that "a billion (i.e., a million million) hypotheses that a fantastic being might guess would account for any given phenomenon."[20] He also had a linked knowledge base in mind, of some sort anyway: "For this phenomena would certainly be more or less connected in the mind of such a being with a million other phenomena (for he would not be restricted to contemporaneous events)." He concluded, in a kind of dismissal, by "not carry[ing] this idea out further," because any such being lacking "nature's pattern" would be faced with an impractical possibility of guessing correctly by chance.[21] Unfortunately, AI can't dismiss the problem; it is the precise problem that must be solved.

THINKING, FAST AND SLOW

The idea that part of our thinking is driven by hardwired instincts has a long pedigree, and it appears in a modern guise with the work of, for instance, Nobel laureate Daniel Kahneman. In his 2011 best seller, *Thinking, Fast and Slow,* Kahneman hypothesized that our thinking minds consist of two primary systems, which he labeled Type 1 and Type 2. Type 1 thinking is fast and reflexive, while Type 2 thinking involves more time-consuming and deliberate computations.[22] The perception of a threat, like a man approaching with a knife on a shadowy street, is a case of Type 1 thinking. Reflexive, instinctual thinking takes over in such situations because (presumably) our Type 2 faculties for careful and deliberate reasoning are too slow to save us. We can't start doing math problems—we need a snap judgment to keep us alive. Type 2 thinking involves tasks like adding numbers, or deciding on a wine to pair with dinner for guests. In cases of potential threat, such Type 2 thinking isn't quickly available, or helpful.

Kahneman argued in *Thinking, Fast and Slow* that many of our mistakes in thinking stem from allowing Type 1 inferences to infect situations where we should be more mindful, cautious, and questioning thinkers. Type 1 has a way of crowding out Type 2, which often leads us into fallacies and biases.

This is all good and true, as far as it goes. But the distinction between Type 1 and Type 2 systems perpetuates an error made by researchers in AI, that conscious intelligent thinking is a kind of deliberate calculation.

In fact, considerations of relevance, the selection problem, and the entire apparatus of knowledge-based inference are implicit in Type 1 and Type 2 thinking. Kahneman's distinction is artificial. If I spot a man walking toward me on a shadowy street in Chicago, I might quickly infer a threat. But the inference (ostensibly a Type 2 concern) happens so quickly that in language, we typically say we perceive a threat, or make a snap judgment of a threat. We saw the threat, we say. And indeed, it will kick off a fight-or-flight response, as Kahneman noted. But it's not literally true that we perceive a threat without thinking. Perceived threats are quick inferences, to be sure, but they're still inferences. They're not just reflexes. (Recall Peirce's azalea.)

Abduction again plays a central role in fast thinking: it's Halloween, say, and we understand that the man who approaches wears a costume and brandishes a fake knife. Or it's Frank, the electrician, walking up the street with his tools (which include a knife), in the shadows because of the power outage. These are abductions, but they happen so quickly that we don't notice that background knowledge comes into play. Our expectations will shape what we believe to be threat or harmless, even when we're thinking fast. We're guessing explanations, in other words, which guides Type 2 thinking, as well. Our brains—our minds, that is—are inference generators.

In other words, all inference (fast or slow) is noetic, or knowledge-based. Our inferential capabilities are enmeshed somehow in relevant facts and bits of knowledge. The question is: How is all this programmed in a machine? As Levesque points out, some field, like classic AI's knowledge representation and reasoning, seems necessary to make progress toward artificial general intelligence. Currently, we know only this: we need a way to perform abductive inference, which requires vast repositories of commonsense knowledge. We don't yet know how to imbue machines with such knowledge, and even if we figure this out someday, we won't know how to implement an abductive inference engine to make use of all the knowledge in real time, in the real world—not, that is, without a major conceptual breakthrough in AI.

THE SELECTION PROBLEM AS UNDERCODED ABDUCTION (AND MORE)

The late novelist, semiotician, and philosopher Umberto Eco, writing in a little-known compilation about abduction and inference, classified types of abductive inference in terms of their inherent novelty (and thus computational difficulty).[23] His classification is instructive for our current concern with AI. Hypotheses or overcoded abductions are, paradigmatically, translation cases, where, for instance, *man* in English means "human adult male." Eco points out that even these seemingly trivial inferences are only partially automatic, because background knowledge and context can alter our beliefs. In a foreign culture with polyglot languages spoken, hearing *man* might not license the dictionary meaning in English. Overcoded abductions still disguise the contribution of a belief or hypothesis. Background belief informs expectations of what phonemes (word sounds) mean.

Undercoded abductions require selection of relevant rules and facts that are already known, but that are applicable to inferences only in context. Understanding natural language is a paradigm case of undercoded abductive inference. *Margaret saw a magpie on the tree. She loathed it* is typically interpreted to mean Margaret loathed the magpie, resolving the pronoun *it* to its preferred antecedent *magpie.* But *Margaret saw a songbird land on her favorite tree. She loved it even more* switches the preferred interpretation to the tree—and nixing the *even more* requires further background about Margaret, the songbird, and the tree. It's undercoded, in other words, requiring an appreciation of the narrative meaning in which the two sentences appear. A system like Siri or Alexa must perform—at minimum—undercoded abductive inferences to generate more contextual and meaningful responses. The two-sentence examples given are but a drop in the bucket with problems getting such systems to work using existing methods.

Abduction gets harder and harder. It soon departs from all known conceptions of automatic inference or computations. Take scientific discovery, or innovation. Human beings invent languages, concepts, and laws to explain the world. This is creative abduction. Creative abductions "leap" to novel conceptual frameworks themselves. Sir Isaac Newton comes to mind. Not only did he extend mathematics to describe instantaneous rates of change on curves (or acceleration), he gave words in English new meanings, to explain physics. *Gravity* used to mean depth and seriousness—as in *gravitas*—and the force of attraction we now call gravity was understood as tendency or purpose. When Newton hypothesized that objects like apples falling to the Earth are governed by gravity, a mathematically describable but invisible force, he also realized it was, or could be, the same force pulling the moon along its orbit—falling around the Earth, in effect—and pulling the oceans up in high tide. The concepts were invented, essentially.

Turing used the mechanical typewriter as a model to explain universal machines, or computers. Typewriters were common knowledge, but they improbably entered into one of the greatest scientific and technological inventions in human history. The question of automating creative abductions like this is obscure, to put it mildly. Yet humans have made them, and still do make them. Watson and Crick's discovery of the structure of DNA is another famous example, as is Einstein's theorizing about relativity. Innumerable lesser-known creative abductions have moved science and innovation along a progressive path since the dawn of human society.

Creative abductions would be less immediately troublesome to AI if they were confined to flashes of brilliance by Newton, Turing, and others. But interesting and important moments in our personal lives are often creative abductions, too. When we reconceptualize our world, for instance, seeing new meaning in everyday happenings, realizing that relationships are more important than money, or have a religious conversion (or traditions or faith lose their hold), we see things through an entirely new lens. We don't just select from among the web of background possibilities, as in solving a puzzle, except in a far more profound sense. Instead, we come to see the world and its happenings and events in an entirely different way.

This creative leap happens daily and for many people. The leaps can be large and serious (as with questions of faith and doubt) or they can be small and mundane. They're often fun. We engage in interesting inference whenever we read the newspaper, or have a conversation, or navigate busy streets to pick up groceries. And creative abductions sit behind our fascination and enjoyment of music, art, film, and stories. Detectives—from the days of Dupin and Sherlock Holmes—entertain because we recognize the creative potential of the human mind in their cogitations. We marvel at how they can extend intelligent inference into feats of reasoning.

In sum, mysterious and wonderful abductive inferences pervade human culture; they are largely what make us human. Dreams of AI might someday capture these leaps automatically, but in the meantime we should admit that we really don't know how to mechanize our experience.

All roads lead here: AI lacks a fundamental theory—a theory of abductive inference. The problem of language understanding exposes this problem unmistakably, and is a central concern of the next section. Let's conclude this section with a few takeaway summary remarks.

First, deductive inference gives us certain knowledge. If the premises in a deductive argument are true, and if the rule used to infer the conclusion is valid (known to be truth-preserving), then deduction guarantees that we move from one true inference to the next. The problem is that little of the everyday world is captured by timeless truths, and even when we have certainty, deductive inference ignores considerations of relevance. Thus, deduction-powered AI can be quite silly and stupid: it might conclude, for example, that a husband won't get pregnant because he took his wife's birth control pills.

Second, inductive inference gives us provisional knowledge, because the future might not resemble the past. (It often doesn't.) Logical experts call induction synthetic because it adds knowledge, but notoriously it can provide no guarantee of truth. It suffers also from inability to capture knowledge-based inferences necessary for intelligence, because it is tied inextricably to data and frequencies of phenomena in data. This gives it a long-tail problem, and raises the very real specter of unlikelihood and exceptions. Inductive systems are also brittle, lacking robustness, and do not acquire genuine understanding from data alone. Induction is not a path to general intelligence.

And third, intelligent thought involves knowledge that outstrips what we can bluntly observe, but it's a mystery how we come to acquire this knowledge, and even further, how we apply the right knowledge to a problem at the right time. Neither deduction nor induction illuminates this core mystery of human intelligence. The abductive inference that Peirce proposed long ago does, but we don't know how to program it.

We are thus not on a path to artificial general intelligence—at least, not yet—in spite of recent proclamations to the contrary. We are still in search of a fundamental theory.

Chapter 13

• • •

INFERENCE AND LANGUAGE I

To talk to Eugene Goostman, you have to chat with him on text. He's not available for a phone call, and you can't have lunch with him. Text him, and he'll tell you he's a thirteen-year-old kid from Odessa, Ukraine. Like a lot of teenagers, his texts are flippant, evasive, overconfident, and prone to misdirection and dissembling. He's rude, then playful. He quips. What he won't tell you is that he's actually a computer program, a chatbot designed by Russian researchers to convince humans he's flesh and blood.

Goostman made history, purportedly, on June 7, 2014, by passing the Turing test, sixty years after Turing's death. In the much-ballyhooed event hosted by the University of Reading in England and conducted at the Royal Society in London, Goostman the chatbot convinced thirty-three percent of selected judges in a five-minute text exchange that he was human.

The event made major news—blogs and news organizations around the world covered it—even though it wasn't a real Turing test. Fully two-thirds of the judges didn't fall for Goostman's tricks, after the test was cut off at five minutes. Still, press coverage of the event, billed as the "at long last" moment, the passing of Turing's challenge to usher in AI, was predictably euphoric. *The Independent* hailed Goostman as a "breakthrough," adding that the program was a "supercomputer."[1] *Time* proclaimed that "The Age of Robots is Here." The BBC called it a "world first"—technically accurate—while the popular tech blog *Gizmodo* informed its readers, "This is big." The press reaction was a testament to the continuing fascination with Alan Turing, no doubt. It was also recognition of the momentousness of really passing his test. Imagine, if you can, really chatting with your computer. The Goostman success dangled this perennial sci-fi dream in front of us.[2]

EUGENE GOOSTMAN IS A FRAUD

There was only one problem. Goostman was a fraud. Not long after the London spectacle, which won the Russian team the $100,000 Loebner Prize for passing the Turing test, computer scientists and commentators began complaining. For one, Goostman didn't really answer questions. He usually changed the topic or blurted out sarcasm. Gary Marcus, then at New York University, wrote in the *New Yorker* that the Goostman performance was little more than a "parlor trick." He echoed Hector Levesque, who also dismissed Goostman-like performances as "cheap tricks." As Marcus explained, the Goostman misdirection tactic creates the illusion of intelligence without requiring any:

> Marcus: Do you read the *New Yorker*?
>
> Goostman: I read a lot of books . . . so many—I don't even know which ones.[3]

These exchanges are like the ones people had with the 1960s chat program ELIZA, which mimicked a Rogerian psychotherapist:

> Patient: Well, I've been having problems with my husband.
>
> ELIZA: Tell me more about your husband

While admittedly fun, they fail to make any headway on the very real natural language challenges facing AI. In fact, Goostman (and

ELIZA) succeed by avoiding language understanding, and for that reason they're irrelevant to a serious conversational test.

Perhaps understandably, pundits and commentators in the wake of the Goostman flap have resorted to decrying the Turing test itself as an illegitimate milestone. *Gizmodo* quickly ran a piece claiming Turing's test was "b*llshit." *Wired* questioned whether the goal of having a fireside chat with a computer should really be the benchmark for full-fledged AI.[4]

The critics have a point, considering the Goostman debacle. Cheating by spitting back sarcastic answers without real understanding hardly measures language ability in the way Turing intended. Cheating at the Turing test by resorting to ploys and cheap tricks exposes a vulnerability we have to lowering the bar for our detection of intelligence when expectations about conversational dialogue are themselves lowered.[5] We might play along with thirteen-year-old Ukrainian boys, thinking they know (and could care) less about adult dialogue. Likewise we might assume in a therapy session with ELIZA that typical therapeutic interaction involves a deliberate attempt to get us to talk—thereby excusing the therapist from doing our thinking for us. Our expectations in these types of situations fit a social context that precludes assessment of intelligence by the respondents in the first place. So it's no wonder that AI researchers have mostly abandoned the Turing test challenge. Stuart Russell's dismissing remark that "mainstream AI researchers have expended almost no effort to pass the Turing test" reflects this frustration with media frenzy about parlor trick performances.[6] It's a weakness in the field to accept them.

But the dismissal is entirely unnecessary. For one, an honest Turing test really is a high-water mark for language understanding. As Ray Kurzweil has pointed out, an alien intelligence might not understand English conversation, but any intelligence that did pass a legitimate Turing test must be intelligent. "The key statement is the converse," he says: "In order to pass the test, you must be intelligent."[7] Kurzweil suggests that future competitions simply allow for a longer test, ensuring that cheap tricks get filtered out in continuing dialogue.

This is just one suggestion. To go further, we might simply exclude the contribution of cheap tricks by adding a rule of play: Contestants could be required to answer questions directly, as if in front of a judge and sworn to "tell the whole truth and nothing but the truth." Expectations in a courtroom certainly preclude ELIZA-like performances. Imagine responding to a prosecutor's or judge's question with "Tell me more about that night yourself. How did that make you feel?" Or we could instruct all contestants to play as if interviewing for the job of understanding conversational English—not a bad test for future voice-activated personal assistants! In such cases, tricks would be immediate violations of conversational rules baked into the test. Goostman would be sunk.

Computational linguists and AI researchers have known all along that engaging in open-ended dialogue is formally more difficult than interpreting monologue, as in understanding a newspaper article. Yet another way to preserve Turing's intuition that natural language ability is a suitable test of human-level intelligence is to consider a simplification of his original test, requiring only monologue. We can do this in the context of a question-answering session, as with the original test. Consider a test simplification: we'll call it the Turing Test Monologue. In a Turing Test Monologue, the judge simply pastes in a news article or other text, then asks questions requiring an understanding of what it says. The respondent must answer those questions accurately. (Goodbye to tricks.) For instance, a judge might paste in the AP article "Your Tacos or Your Life!" and ask the respondent whether the story is funny or not, and why? Passing this test would be, strictly speaking, a logical subset of a completely open-ended Turing test, so it would be entirely fair to use it—in fact, it would give advantage to the machine, which might not completely understand how to handle "pragmatic phenomena" in back-and-forth dialogue—more on this later.

Unfortunately, with the Turing Test Monologue, we already have evidence that AI systems are hopelessly lost. Gary Marcus and Ernest Davis point out that the state of the art for reading comprehension by machines is pitiful. Microsoft and Alibaba were much celebrated by the media for increasing a baseline score on a reading test known as the Stanford Question Answering Dataset (SQuAD), but only when the complete answers were in the text. This was therefore a simplified task of "underlining" answers as explicitly provided, cued by questions that clearly pointed to them.[8] Marcus and Davis go on to highlight embarrassing performances on seemingly simple questions, such as one asking only for the name of the quarterback mentioned in a snippet of Super Bowl coverage. The failure of understanding in such cases is obvious. So even when we eliminate trickery, AI language understanding is in trouble. The Turing test remains a legitimate assessment, whose bar is, if anything, set too high.

The more we take a serious look at the requirements of language understanding, the more daunting passing even the simplified Turing Test Monologue becomes. Hector Levesque devised a vastly simplified version of the test and called his quizzes Winograd schemas (after AI pioneer Terry Winograd, who worked on natural language understanding). Winograd schemas require answers to multiple-choice questions about the meaning of single sentences in English. That's a far cry from the Turing test. And yet, AI researchers are a far cry from mastering them.

THE CURIOUS CASE OF WINOGRAD SCHEMAS

Hector Levesque is one of the few AI scientists today still focused on knowledge representation and reasoning. Admirably, Levesque wants to imbue AI programs today with more than statistical techniques for analyzing big data: he wants to give them common sense.

Levesque proposed a simplified version of the Turing test which is much easier than the original open-ended and unrestricted test, but that, importantly, still frustrates all known automated approaches to language understanding. In 2013, Levesque presented a paper, "On Our Best Behavior," to the *Internet Joint Center for Artificial Intelligence,* which was quickly recognized as a call to arms for genuine AI.[9] Inspired by the full Turing test, Levesque suggested we pose questions to machines that require some deeper understanding of what's being said. The test questions are single sentences, not entire conversations. For example: *Can a crocodile run a steeplechase?* Levesque deliberately chooses examples like this because non-expert humans, having ordinary common sense, can figure them out (crocodiles cannot run a steeplechase), but popular tricks like using a search engine to look up the answer won't work. Since there won't (one assumes) be any web pages discussing crocodiles running steeplechases, there will be no way to bypass the need for understanding. AI systems run aground on such examples; an answer occurs to humans almost immediately.

A Winograd schema is a multiple-choice exercise, which removes the possibility that the machine can resort to misdirection, sarcasm, jokes, or the appearance of a bad mood to bluff human judges in any situation where a direct answer would reveal its lack of understanding. The schemas are based on a common feature of natural language, as Winograd's original question inspiring them makes clear:

> *The town councilors refused to give the angry demonstrators a permit because they feared violence. Who feared violence?*
> a) The town councilors
> b) The angry demonstrators

Note the pronoun *they.* It's a plural pronoun; it might refer to the councilors or the demonstrators. It's ambiguous, in other words, because either answer is possible without breaking the rules of grammar. Yet only one really makes sense. Humans get the right answer to questions like this effortlessly and with near hundred-percent accuracy. AI systems do not. Their performance on Winograd schemas is not much better than random guessing.[10]

Working with other AI researchers, Levesque exhumed Winograd's challenge in 2012, when big data and machine learning clearly dominated approaches to AI (as they still do). He gathered a test set of multiple-choice questions, all exploiting the common and ubiquitous feature of ambiguity in natural language. He called the ambiguity challenge the pronoun disambiguation problem. It captured the inspiration behind the Turing test, but in a simplified form: ordinary natural language understanding, like English or French (or what have you), requires general intelligence. In particular, Levesque believed that AI systems would require knowledge about what the words in language actually mean to do well on the test. Here's another example of a Winograd schema:

Joan made sure to thank Susan for all the help she had given. Who had given the help?
a) Joan
b) Susan

To insulate the test from the problem of cheap tricks used to fool judges on the Turing test, Levesque added a twist: two words designated as special that "flip" the answer, while leaving the rest of the question unchanged. In this example, the special words are *given* and *received*. Exchanging these designated special words generates another question:

Joan made sure to thank Susan for all the help she had received. Who had received the help?
a) Joan
b) Susan

Here's another schema, employing the special words *golfers* and *dogs*:

> *Sam tried to paint a picture of shepherds with sheep, but they ended up looking more like golfers. Who looked like golfers?*
> a) The shepherds
> b) The sheep

Winograd schemas are vastly simplified compared to the original, conversational Turing test, but by posing multiple-choice questions that require resolving pronoun reference (pronoun disambiguation), they capture what earlier researchers called "common sense holism"—the idea that natural language cannot be understood by dissecting sentences, but requires a general understanding. Thus Winograd schemas are simple but typically unusual questions that make perfect sense to readers that have ordinary knowledge about the world. To take Marcus's example, with basic knowledge of alligators and high hedges, it's clear that the short legs of an alligator disqualify it from competing by clearing hurdles. Why is this a difficult problem for AI?

In part, schemas are difficult because the two referents—the choice of nouns and noun phrases like *alligator* and *hurdles*—rarely (if ever) occur together in web pages and other text. Since data approaches to AI rely on lots of examples to analyze statistically, odd questions like those found in Winograd schemas represent a significant challenge—all in one sentence. In fact, Winograd schemas are quite bulletproof to tricks like counting web pages. They're "Google-proof," as Levesque puts it. But the requirement of ordinary knowledge for their interpretation is a still deeper reason computers don't perform well on them. Changing Marcus's subject from *alligators* to *gazelles* would change the answer (gazelles can leap hurdles, no problem), but again the question is odd and thus quite rare on the web. Machine learning and big data don't help. AI systems can't look up the answer.

As we might expect, AI researchers have devised new tricks to tackle Winograd schemas. Because so much content is available for analysis on the web today, frequency patterns can still be exploited in some cases. For instance, to answer questions about whether *x* is taller than *y*, researchers can mimic real knowledge by performing web searches with the pattern "*x* is taller than *y*." If the people, buildings, or whatever things standing in for *x* and *y* happen to be in repositories like Wikipedia, data can be found, and math employed to answer the question—all by processing info-fields and other data sources on the web. But, again, Levesque anticipated this technique and suggested examples that use common nouns (*gazelles, alligators, pens, paper, bowling balls, sheep*, and so forth) that don't appear in curated databases or online encyclopedias. This stymies the search-the-web trick, while also highlighting the commonsense, ordinary quality of knowledge required—preserving Turing's original intent to use basic conversation that everyone engages in day to day.

Winograd schemas also guard against search-engine tricks in another important way. The relation between *x* and *y* can be modified too, just like the things discussed (*alligators* versus *gazelles*). This is another roadblock to using data tricks, all the while preserving the simplicity of the questions. Consider first a schema requiring knowledge of the relative sizes of common objects:

> *The trophy would not fit in the brown suitcase because it was so small. What was so small?*
> a) The trophy
> b) The brown suitcase

There is a trick—a technique—for this schema, actually. As Levesque puts it, in quasi "computer-ese" language, we can dissect the question into a relation (call it *R*), and a single property (call it *P*):

> *R* = does not fit in
> *P* = is so small

Then we use "big data," he says, and "search all the text on the web to determine which is the more common pattern: 'x does not fit in y' + 'x is so small' *vs.* 'x does not fit in y' + 'y is so small.'"[11] For the scores of example sentences returned by search, a pattern might emerge where, for instance, the second item mentioned is more frequently smaller than the first. Imagine a social media post in which someone is complaining about packing and keeps mentioning that such-and-such won't fit because their backpack is too small. Simply counting search results for the two patterns can yield a statistical answer, and if it's correct, a system can answer Winograd schema questions without knowledge.

Unfortunately, the approach is quite shallow. It is powerless against even small modifications that change the meaning of the question, and hence switch the preferred answer. Our original "relation *R*" trick fails, for instance, on this question:

> *The trophy would not fit in the brown suitcase despite the fact that it was so small. What was so small?*
>
> a) The trophy
> b) The brown suitcase

It's the trophy that's so small now, not the suitcase. Thus the specification of *R* and *P* will generate an incorrect response. The modified question is probably less frequent than its original version, so big data is a hindrance, not a help. This is the rub. Even with simple, one-sentence questions, the meanings of words—*alligators, hurdles*—and the meanings of the relations between things—*smaller than*—frustrate techniques relying on data and frequency without understanding. Winograd schemas are a window into the much greater difficulty of the Turing test.

Machine learning and big data have made significant progress on some problems in the last decade. As a rule, though, workarounds that sidestep actual knowledge and understanding account for the successes. There is an ongoing confusion here, particularly with automated language translation systems which seem to require some sort of language understanding; take Google Translate, which is often touted as a runaway success story proving how AI is quickly taming natural language. But the relatively recent availability of large volumes of translated texts on the web facilitates data-intensive approaches, which deliver "good enough" results by exploiting mappings between words and sentences in texts translated into different languages. The mappings are mostly all in the data, so inductive strategies work. Understanding is not required.

For instance, official Canadian parliamentary documents are translated word for word from English into French; in a case like this, approaches like deep learning suffice to "learn" the mappings between the languages. No actual understanding of English or French is required. For example, if you type *John met Mary at the café* into Google Translate, and ask for a translation to French, you'll get *John a rencontré Mary au café*. This is a perfectly acceptable translation. Other examples fail, however. The failures typically involve ambiguity somewhere, as with referential phenomena like pronouns, or polysemous (many-meanings) words embedded in context. (If you want to experiment with Google Translate on sentences containing ambiguity, try *The box is in the pen* or *I loved the river. I walked to the bank.*)

Confusingly, work in the 1960s on so-called fully-automated high-quality machine translation began with statistical approaches, albeit simpler ones. They didn't work very well, and soon apostates like Yehoshua Bar-Hillel concluded that automatic translation was hopeless, because knowledge and an appreciation of context seemed necessary—reasons that later inspired Winograd and then Levesque to formulate knowledge-based schemas. But Bar-Hillel didn't anticipate the contribution that big data would make a few decades later.

Statistical approaches started showing promise in the 1980s, with work by IBM researchers, and web giants like Google used similar techniques (and now deep learning) to provide decent statistically-driven translation services. Fully-automated high-quality machine translation came full circle.

But Bar-Hillel's original skepticism is still germane; translations requiring knowledge and context are still a black box to modern approaches, big data and all. For instance (and ironically), Google Translate as of October 2020 still gets Bar-Hillel's 1960s example wrong. Bar-Hillel asked how to program a machine to translate *The box is in the pen* correctly. Here, *pen* is ambiguous. It might mean a writing instrument, or it might mean a small enclosure for animals. Using Google Translate, *The box is in the pen* translates to *La boîte est dans le stylo,* in French, where *stylo* means writing instrument, which isn't the preferred interpretation (because boxes are typically larger than writing pens). In other words, "good enough" translation depends on one's requirements. Google Translate might get less contextualized sentences most of the time, but there will be errors—and the errors might matter, depending on the person using the service. Services like Google Translate actually underscore the long-tail problem of statistical, or inductive, approaches that get worse on less likely examples or interpretations. This is yet another way of saying that likelihood is not the same as genuine understanding. It's not even in the same conceptual space. Here again, the frequency assumption represents real limitations to getting to artificial general intelligence.

"Good enough" results from modern AI are themselves a kind of trick, an illusion that masks the need for understanding when reading or conversing. On image-recognition tasks, systems might be forgiven for incorrectly classifying certain images (but not if the system is a self-driving car). But on language tests like schemas, a 20 percent error rate (thus 80 percent accuracy) means that fully two out of ten examples are opaque to the system. And accuracy rates on Winograd schema tests are much worse—barely better than random. As schema questions make clear, errors are less excusable when they are simple sentences, with no inherent difficulty for humans.

The point is that accuracy is itself contextual, and on tests that expose the absence of any understanding, getting six out of ten answers correct (as with state-of-the-art systems) isn't progress at all. It's evidence of disguised idiocy. Like the mythical androids in the sci-fi thriller *Blade Runner,* a test looking for mind (or real emotion) will eventually unmask machines as programmed impostors. Only, the Winograd schemas are short and quite simple. All it takes is one question, requiring some basic knowledge of what's being said.

Levesque's "pronoun disambiguation problem" is a vastly simplified step in a very large minefield that automated systems will have to navigate to have a prayer of managing ordinary conversations. Resolving pronoun references is just one knowledge-based problem among a plethora of others that must be solved for genuine language understanding. And Winograd schemas hugely simplify in other ways, as well. In addition to the one-sentence limit for each schema, the test also simplifies by ignoring most of the pragmatic phenomena in conversational dialogue. Pragmatics is hell for automation. To see why, we'll next tour through the field of natural language processing, or understanding.

Chapter 14

• • •

INFERENCE AND LANGUAGE II

Natural language is a complicated beast. Its ingredients are symbols, letters, and punctuation, forming words: w-o-r-d-s. Words are chunks of meaning. They refer to all sorts of things: objects like the *razor,* feelings like *pleasant,* moral judgments like *bad,* situations or events like *party* or *elections,* and abstract collections like the *economy* or the *farm.* Words, having meaning, can also be ambiguous: the *farm* might mean the physical location of a farm, or it might mean the business of the farm, or the place of living for a family. The exact meaning of even a simple word like *farm* will depend on what the speaker really intends, which is contextual.

Understanding context expands the scope of language interpretation to surrounding words, sentences, paragraphs, and even entire texts. Ambiguities must be resolved in phenomena like pronouns and indexicals—where and when ("at the appointed time," when context makes clear that midnight is meant). Finally, we have what linguists call pragmatics: the context that includes the person making the statements, his or her purposes, interests, and the like. Sarcasm, irony, and other aspects of communication come into play. Language starts with letters and words, and ends up with questions about meaning and mind.

A TOUR OF LANGUAGE UNDERSTANDING

Language understanding (say, understanding a tweet, a blog, or *War and Peace*) is a kind of inverted pyramid, like the old food pyramid illustrations used to depict a healthy diet, but stood on its head. As you go up the (inverted) pyramid, more meaning is encountered, and computational treatments of language run into increasing difficulties. At the bottom, the tip, we have orthography, the letters and punctuation that combine into words and sentences. Then lexical and morphological features like words, their prefixes, suffixes, parts of speech, and so on. To form sentences we need words combined into phrases, like the noun phrase *the ferocious lion*, which contains a subject and some identifying and qualifying information; then we must add some action, like *The ferocious lion devours*; then we must add an object if that action (and verb) is transitive, like *The ferocious lion devours meat.* Sentences form paragraphs, which express a topic and flesh it out, and multiple paragraphs joined together form a story or narrative. The story (of one or more paragraphs) is called a discourse, or just a text. An example of a large text is a book, which has a genre, like fiction or nonfiction, and so on. A transcript of a conversation is also a text. Thus the Turing test sits at the top of an inverted pyramid of language, where it's wider and expanded into a full dialogue, a conversational text.

In some sense, the inverted pyramid of language can be viewed purely syntactically, as the construction of a text using rules that specify how to combine characters into words, words into sentences, and sentences into paragraphs. Syntax can be processed and analyzed by a computer. But the language pyramid leads a double life—it's also about meaning. Natural languages use symbols and rules to convey meaning, which can't be ignored if we want to say anything useful using language at all. Language is a journey into meaning. A children's book takes this same journey as a Russian novel. It's all the expression of meaning.

In linguistics and other cognate fields, semantics is the study of what things mean. Semantics means "meaning," and semantic analysis is roughly what's involved in taking multiple-choice Winograd schema tests. Knowledge and belief is required, not only syntax and information about symbols. Inference is required to apply knowledge in context. If I believe, for instance, that an alligator can run the hundred-meter hurdles, I have to explain how its little legs can manage the required jumps over the hurdles, and to do this I have to know something about the anatomy of an alligator. (Not much explanation is required for humans, actually: a picture will do, as any five-year-old has probably seen.) The answer to the question is not in the syntax. It's in its semantics.

Winograd schemas replace the syntactic pyramid of language with a semantic one, where words have meanings, referring as they do to objects in the world. But words in sentences sometimes make sense only in the context of an entire text. Text interpretation invariably involves pragmatic considerations, which add to word meanings inferences about purpose and intent. A conversation between two people is an example of pragmatics in language understanding. If I ask "Can you pass the salt?" and you reply "Yes" but do nothing, then the context—my intention—escapes you (or you jest—which is more pragmatics). In pragmatics, what is said is tightly dependent on how it is said and why it is said.[1] What people mean is almost never a literal function of what they word-for-word say. This feature of ordinary talk, studied in linguistics as pragmatics, is what makes language interpretation hard for AI, but meaningful and interesting—and generally easy and natural—for people.

Syntax remains essential. If we switch languages from English to, say, Chinese, then we also switch from Latin characters to pictograms. If someone doesn't speak or read Chinese, then attempts at communication are futile. But, as logicians might say, knowledge of syntax is necessary but not sufficient. World knowledge is essential. If I tell you that someone was hit with a Styrofoam bat, it matters what you know about Styrofoam, because that piece of knowledge determines what you'll think about my comment, and how you'll react: "Tragic," you might say, sarcastically. But an AI system that assumes Styrofoam must be a good, hard substance for swinging a bat might be rather alarmed: "Is she okay?" Even the simplest exchange typically requires real knowledge and context-dependent inferences, involving semantics (and pragmatics).

Turing understood this in his original exposition of the prospects for AI. In his example, he imagines a human questioner (the interrogator) and a computer (the witness). The interrogator has asked the witness to write a sonnet, who reproduces Shakespeare's Sonnet 18:

> Interrogator: In the first line of your sonnet which reads "Shall I compare thee to a summer's day," would not "a spring day" do as well or better?
>
> Witness: It wouldn't scan.
>
> Interrogator: How about "a winter's day," that would scan all right.
>
> Witness: Yes, but nobody wants to be compared to a winter's day.
>
> Interrogator: Would you say Mr. Pickwick reminded you of Christmas?
>
> Witness: In a way.
>
> Interrogator: Yet Christmas is a winter's day, and I do not think that Mr. Pickwick would mind the comparison.

Witness: I don't think you're serious. By a winter's day one means a typical winter's day, rather than a special one like Christmas.[2]

The interrogator's first question tests the computer's knowledge of sonnet writing, specialized knowledge that we might forgive many perfectly knowledgeable people for lacking. Iambic pentameter excludes "spring" day, and the computer curtly replies that the interrogator's suggestion won't "scan."

"Winter's day" does, though, so why not that? Here the focus shifts to the metaphor of comparing one's love to a winter's day. The witness brushes off the interrogator's suggestion of a "winter's day." We can assume winter days are short and cold, and someone deep in romantic love would not choose such imagery. But the shift from summer to winter is not so obvious, actually. Winter days can be beautiful and snowy, with brilliant sun illuminating the soft silence of forests. The idea is rather that, in love, if we are to pick a season to describe our beloved, the warm, long, beautiful days of summer make more sense.

The interrogator then suggests a "special" winter's day, Christmas. This tests the computer's knowledge of the intent or purpose of the sonnet, which is not to compare someone to a celebration, but to something lovely and "constant," something like the experience of a lovely day in summer. To compare one's love to Christmas, or to the Fourth of July, is to mix together ideas and emotions that entirely blunt and ruin the point of the sonnet. The witness, apparently aware of the foolish misdirection, shoots back that it doesn't think the interrogator "is serious."

Understanding the point of the sonnet, in other words, is to understand the deep emotion of someone in love comparing his beloved to a summer's day, something beautiful and long and lovely, though the poet believes that his love is the more so, made immortal by his deep passion and by the timelessness of the sonnet he's written. The exchange between the interrogator and the witness requires knowledge, to be sure, but most of the relevant knowledge is in fact about people—people in love and how we express that in words and feelings. To understand the sonnet we have to know what it's like to talk about a beloved. The interrogation is about mental states and emotions. This is pragmatics, in its broad and intended sense.

Poetry, we might say, is intrinsically pragmatic. But less effusive prose still showcases the ubiquity of pragmatic phenomena in language, in nearly every "ordinary" sentence. It's inescapable. *Deixis,* for instance, refers to words like pronouns that cannot be understood or disambiguated without knowledge of context, like *me* or *he* or *here.* Deixis means "pointing," and language often points: *She saw a sparrow here last night* points to whoever saw the sparrow (person deixis), understood by the listener in context. Same, too, for *here,* referring to some location in context (spatial deixis), and *last night,* which must refer to the night before the comment (temporal deixis). A system engaged in a conversation, not resorting to canned tricks, must keep track of all this.

At minimum, deixis is just one problem among many. Language is full of contextual subtleties that require a deeper understanding of intent. If I mention that I was playing Mozart last night, the comment is naturally adjusted to make sense: I was playing music written by Mozart last night. (I wasn't "playing" the man Mozart, in some urban-dictionary sense of talking him out of his wallet, say.) We usually don't bother with such adjustments because pragmatic phenomena in language are part and parcel of normal communication. We infer intended meaning, not by seeing lots of examples but by thinking through what is meant.

Linguists (and the rest of us) know that people often mean much more than they say. This reveals itself in language in different ways.

Conversation is full of ellipses, for instance, where nouns, verbs, and other parts of speech and phrases are entirely left out: *I went to the mall yesterday, and Shana went today.* We possess prior knowledge of specific people, places, and things, and use this knowledge to shorten and make more natural our communication with each other. *Can you lend me your book?* assumes you have a book, and also that you know which one I mean. If you inform me that *Charles stopped studying mathematics when he won the poetry contest,* I'm presupposing that you know Charles, and that Charles was studying math, and so on. Language, we might say, is partially occluded: we leave out details and assumptions; we count on other minds to infer and assume what we mean.

Anaphora, mentioned earlier, means "referring back," as in *The ship left the harbor in May. Roger was on it.* Here, the pronoun *it* refers back to *the ship.*[3] Anaphora (and its cousin, cataphora, or "referring forward"), as we've seen, is introduced into language with placeholders, typically pronouns like *he, she, they, them, it,* and so on. Pronouns aren't particularly rare, and anaphors in general are ubiquitous. They are one example of language pointing to itself, and outside itself into the world.

Anaphors also interact with other pragmatic inferences to convey meaning, as in *We loved the quartet last night. It was all so lovely,* which subtly shifts the otherwise anaphoric *It* (referring to the quartet) to the evening itself. Anaphora itself is a special—though common—case of reference in language, of which a popular type in computational treatments of language is known as co-reference, where two mentions (words or phrases mentioning something) in a discourse or text refer to the same object or situation in the world. In the example above, *The ship* and *it* co-refer to the ship that left the harbor in May. (If Roger was *in* it instead of *on* it, we can imagine scenarios where he was in the harbor: *The ship almost drowned him as it passed by.*) Winograd schemas in effect simplify the general, computational co-reference problem into a single sentence (rather than an entire text, say) that always contains two possible antecedents: nouns or noun phrases in the sentence, where one and only one co-refers with the pronoun:

> *The sack of potatoes had been placed below the bag of flour, so it had to be moved first. What had to be moved first?*
> a) The sack of potatoes
> b) The bag of flour

The schema is a co-reference problem; *it* refers to the same thing in the world as the noun phrase *the bag of flour* (they both refer to a bag of flour, the one above the sack of potatoes described in the sentence). But schemas do simplify, quite a lot. For one thing, they eliminate open-ended reference: *Later that year, imports fell precipitously. By year end, the entire economy was in trouble.* Here, *The entire economy* refers (to the entire economy) but without an attachment in the text itself, as with schemas. (In this example, there is an implicit meronymic, or part-whole relationship between imports and the whole economy.) By contrast, Winograd schemas always offer the interpreter two choices, which are clearly specified.

Conversations, op-eds in the newspaper, and a letter to a friend present much more complicated interpretation tasks. Part of the schema simplification is structural: because we know, taking the test, that there are only two possible answers, a strategy of randomly guessing won't be so bad, around 50 percent accuracy. Not bad—and with zero intelligence! But humans score 95 percent or better at Winograd schemas, and the best systems to date don't perform much better than a simple heuristic that picks the first answer, or randomly guesses.[4] The anaphors are an explicit knowledge and inference condition on interpretation—not a test of frequencies in data. The conclusion is particularly devastating because there's nothing "tricky" about the test otherwise. The questions are fully articulated, and the choices are clear. For the most part, people find them easy (if a bit odd).

Consider a more complicated language test, the Turing Test Monologue described above, where the AI system must read a snippet of monologue, like a story or a news article, and answer questions about it posed by a human judge. Here's an example test story:

> A man moved to a new town after he quit his job, hoping to start a new life. He drove to the store to see if his employee discount still worked. At the self-checkout, he discovered that it didn't. Muttering to himself, he said "I'm never coming here again." A woman overheard him and gave him a funny look. As he was leaving, he bumped into his wife, and told her what happened. He said again "I'm never coming here again!" She replied, smiling, "Then I will. It's close to our house, and cheap." He nodded, smiling, and said "Fair enough." Later they went to the park, and he compared her to a summer's day.

The story is contrived (I made it up), but it's not particularly difficult for English readers to understand, and it's not unique in requiring commonsense knowledge and pragmatic understanding to interpret it correctly. Let's say a new AI system, DeepRead, using the best machine learning and big data available, has just been released for demo by the hot new startup (sure to be acquired by Google), Ultra++. The human judge asks DeepRead some simple questions:

(1) Did the man walk to the new town?
(2) Why was the man hoping to start a new life?
(3) Did the man die?
(4) Do you think the man drove a car, or a golf cart? A bus?
(5) Did the man used to work in that store? Another store owned by the same company?
(6) Did the man's employee discount still work?

(7) Did the man mutter that he would never come to the store again? The town? Only the space in front of the self-checkout?
(8) Was the woman who overheard him muttering his wife?
(9) What did he mean to communicate to his wife when he bumped into her?
(10) What did she mean by "I will"?
(11) Was the store next door to them?
(12) Did the man love his wife?
(13) Do you think he started a new life, after all?

To have a prayer of answering these questions, DeepRead will have to solve a number of difficult problems on the language pyramid involving semantics and pragmatics, none of which are strictly in the data, in the sense required for data-centric AI.

Start with some commonsense knowledge that *moving to a new town* typically involves packing up and driving or flying to the town. Question (2) is strictly speaking not answerable, but given the context, most human readers would assume his *new life* is connected to his recent unemployment (although "I have no idea" might be reasonable for less adventurous readers). Question (3) also requires understanding that the idiomatic phrase *starting a new life* means you don't first physically die. Question (4) is an example of implicature, or what is implied by a sentence or phrase but not directly stated. We assume the man drove a car—although it's logically possible he got there by golf cart. In the remaining questions, DeepRead must use background knowledge to realize that the man's employee discount once did work, and so the man presumably worked for the company that owns the store he is in (but in a different location—presumably in the town he left); resolve the pronoun *it* to the man's employee discount (anaphora resolution); understand the location deixis *here* to mean the store (not the town or the spot in front of the self-checkout); understand that *a woman* cannot mean his wife (another implicature);

understand that the man informed his wife not only of his intention not to come to the store again, but also that he was frustrated by the company—and the list of difficult language problems grows.

His wife uttering *"Then I will"* is incomplete, involving ellipsis, where a statement is shortened because of assumed knowledge between speaker and hearer. DeepRead should understand that his wife will come to the store again, which is not said. More anaphora: *It* uttered by his wife means *the store,* and her comments about it being close and cheap are reasons why she will shop there. DeepRead should also get that the store is not next to their new house, since the man drove. And finally, that the man's smile meant he was letting go of his frustration, indicated even more by his going to the park later with his wife and comparing her to a summer's day. DeepRead should conclude that the man loves his wife and respects her opinion (he was likely glad to have bumped into her), and that comparing her to a summer's day was an expression of happiness and appreciation.

Alas, I've left out all sorts of required knowledge and inference. DeepRead should understand that the man went into the store, and didn't just drive to it, though this is implied rather than stated. And so on. No existing or foreseeable AI system can answer these questions—and we are only talking about the Turing Test Monologue, using this one simple story. DeepRead doesn't exist. The mind enters into language and pieces together a picture of what is happening, and why. Our inverted pyramid broadens into problems that require a deep and meaningful picture of the world, and entering into the mind of the speaker (or writer) is part of the challenge. This brings us back to pragmatics, and our friend Goostman.

GRICE'S MAXIMS FOR GOOD CONVERSATIONS

Eugene Goostman's designers, you might say, used pragmatics as a tool, a weapon. The chatbot's code exploits an area of pragmatics that turns language understanding on its head, with overuse of sarcasm, full-time dissembling, and copious misdirection. It exploits, in other words, an assumed understanding among judges that thirteen-year-old Ukrainian boys sometimes avoid questions not just because they don't know answers, but because they do not care to try.

Linguists will recognize that Goostman cheerfully violates what are called "Grice's Maxims."[5] In the early part of the twentieth century, the philosopher of language Paul Grice offered four maxims for successful conversation:

1. The maxim of quantity. Try to be as informative as you possibly can, and give as much information as is needed, but no more.
2. The maxim of quality. Try to be truthful, and don't give information that is false or that is not supported by evidence.
3. The maxim of relation. Try to be relevant, and say things that are pertinent to the discussion.
4. The maxim of manner. Try to be as clear, as brief, and as orderly as you can, and avoid obscurity and ambiguity.

Eugene Goostman is a repeat offender. It violates all Grice's maxims, throwing off hapless human judges presumably attempting to ascertain ordinary understanding and a baseline skill and ability in conversation. Well-minded people typically adjust to deficits stemming from English as a second language or from youthful cheekiness, as long as there's some noticeable effort to be informative, sincere, relevant, and reasonably clear. Not so Goostman, who violates the maxims by design.

But violating Grice's maxims in everyday conversation tells ordinary interlocutors that something is "off." If someone approaches you in a coffee shop and says, "I am Brian Johnson, could you tell me what time it is," the request, while understandable, will seem strange. Grice's maxims explain why: the request violates the maxims of quantity and relation, and prompts confusion over information not necessary for getting the time. (We wonder, "Why would he say his name? Does he think he needs to show credentials for asking for the time?")

Back to Goostman: In a structured, time-limited test, with the world watching, it's simply ridiculous to violate Grice's maxims, systematically, when the entire point of the test is to determine intelligence by ordinary conversation. Yet that worked, in the competition anyway. It shouldn't have. And the backlash to rid ourselves of Turing tests was unnecessary and silly, too. To get rid of Goostman, take your pick: use simplified multiple-choice tests like Winograd schemas, or apply fixes to conversational (or monologue-based) tests to place a heavy penalty (like disqualification) on avoidance techniques. Tricks are easy to banish, in other words. The hard problem is getting to understanding.

WHEN GOOD ENOUGH ISN'T GOOD ENOUGH

Poor performances on Winograd schema tests highlight another major weakness of data-driven, modern approaches to AI. In a section of his paper titled "The Lure of Statistics," Levesque said the "data simulation" approach was like trying to do something "vaguely *X*-ish" as opposed to actually doing *X*, where *X* is "one of the many instances of intelligent behavior." This, he insists, is "the overarching question for the science of AI."[6] But Levesque could have pushed his critique even further. Doing something vaguely "*X*-ish," in other words simulating intelligence for some task *X*, invariably pushes difficult problems requiring real understanding into the realm of the unsolved, the few percentage points still not conquered (or, in other words, up the language pyramid of meaning). This becomes a hiding place for problems requiring real understanding. It creates general confusion, too.

The public, witnessing apparently impressive performances, understandably conclude that the remaining performance gaps between machines and humans will melt away as more powerful computing and analysis techniques (and more data) help future systems scale to better performance. But blind confidence that unsolved problems will someday be solved ignores the real distribution of problems and their solutions, with easy ones occurring frequently in data-driven simulations, and other, rarer and more difficult ones falling outside the scope of frequency-based analysis. This means the harder problems—the missed answers—might require an entirely different approach, and not more data. The requisite new approach, as I've argued, presupposes a non-deductive or inductive underlying inference framework. In fact, if DeepRead could perform reliable commonsense abductive inference, it would conjecture correctly about the missing and assumed knowledge and events in the story above. It can't deduce them (they are not logic problems), and as they are not inferred from the syntax (or data) of the story, it can't induce them, either. DeepRead awaits progress on inference; unfortunately, the required inferences are not at present programmable. There is nothing to scale up to—the inference required is distinct, and represents a future conceptual discovery.

I've focused on Winograd schemas because they so clearly require non-inductive inferences while posing what seems to human minds a simple, one-sentence test of understanding. It's hard to debate the unfairness of a test which is only a simple question in English. Winograd schema problems are deliberately about common objects, which means that any one noun or noun phrase will typically occur in texts, like web pages (for instance, *alligator* or *trophy*). No technical knowledge is required. But any two nouns or phrases occurring in a single question drops the expected frequency significantly, in some cases to zero, as in the case of alligators and hundred-meter hurdles. Thus, all the examples are relatively rare in big data, though straightforward. And in cases where the two nouns or noun phrases (as with the trophy and suitcase question) might occur together in a sentence on the web, simply changing the relation between the two nouns returns us to infrequency—and defeats modern data-centric tactics, as we've seen. Hence, for all practical purposes, Winograd schema questions can't be simulated, which explains the poor performance of systems attempting to automate the test.[7]

It follows from the inference framework I've described above that deductive and inductive systems are inadequate for artificial general intelligence. I've also explained that any given type of inference can't be reduced to another (recall the syllogism examples), so that we can't, for instance, call abduction a species of induction with suitable expansions, which would convert it into a completely different symbolic form (or "logical form"). A question remains whether some combination of induction and deduction might scale up to abductive inference. It can't, for the same reason that we can't subsume one type into another—they are distinct and involve fundamentally different abilities. (As a comparison, consider: if I know German and Spanish, might I somehow put these together and understand Russian?) But types of inferences are sometimes combined in work on AI, with interesting results.

Teams of researchers tackling difficult problems like language processing often build large hybrid systems, using specially designed architectures, databases, and algorithms for prediction or inference. Such systems invariably include machine learning as a component or subsystem. They also make use of knowledge bases and inference techniques that hark back to the premodern era of classical AI. Hybrid systems can achieve impressive results. Not playing favorites, they make use of all available techniques, methods, and algorithms to solve difficult problems. One of these systems in particular galvanized the media and public, just as AI itself was emerging as the hot topic in a rapidly changing world. For, at first blush anyway, the system seemed to understand—really understand—how to answer questions posed in English.

WATSON, MY DEAR WATSON

IBM has a knack for engineering AI systems that play games and generate marketing buzz for the legendary behemoth tech company. In 1997, IBM made headlines with Deep Blue, the chess-playing supercomputer that vanquished then-champion Garry Kasparov in a televised and widely anticipated event. Deep Blue was a sensation, in large part because the crossover point, where machines outplayed humans in chess, was supposed to be years away. Pundits speculated that true intelligence had arrived in supercomputers. Talk of the Turing test resumed. Media commentators worried about how long it would be before machines overtook us everywhere else.

In retrospect, the Deep Blue spectacle had little to say about machine intelligence, though it anticipated the spate of data-driven systems that would outplay humans at other games in the coming decades. In the end, chess is a game played according to deterministic rules. Garry Kasparov is a genius, sure, but Deep Blue outplayed him with essentially sheer computational power: evaluating more moves, seeing deeper into the branching game tree. AI enthusiasts over the decades have bemoaned the public's tendency to quickly dismiss new successes as "not really intelligence" after an initial buzz, and Deep Blue was no exception. By the turn of the century it was largely forgotten, another milestone in AI that failed to awe a public who perhaps sensed deep down that chess-playing wasn't a proxy for general intelligence after all. Deep Blue was a showcase for IBM's well-funded engineers and fast computers.

After the internet bubble burst in 2001, excitement about AI temporarily faded. Billions of dollars of investment had evaporated in schemes and visions that failed to see basic realities. AI seemed futuristic and financially insecure, which is precisely what wasn't wanted by investors and entrepreneurs still licking their wounds. The winter was short-lived. By 2004, Google had come into its own, and social networks were on the way. (A few early attempts, like Friendster, failed but the exciting concept was in the air.) Web 2.0 was coming. In 2004, too, IBM's leadership had started looking for another public relations boon. Chess was yesterday's news, but the web was taking shape and reinvigorating everyone for exciting possibilities. Games still attracted attention, and they had the additional benefit for tech companies that AI could often succeed in the artificial constraints of a game, even as it foundered on simple common sense. As luck would have it, one popular game in particular was already in the news: *Jeopardy!*

The television quiz show *Jeopardy!* is a kind of gamified version of a simplified Turing test (or so it seems). It's a broad test of knowledge of facts, with the "conversation" reversed so that an answer is given to contestants, who then must conjure the correct question. For example, to the prompt "Developed by IBM, it beat Kasparov at chess," the correct response would be "What is Deep Blue?" In 2004, *Jeopardy!* ratings were soaring because of a human, Ken Jennings, the returning world champion who had won a record-breaking seventy-four *Jeopardy!* contests in a row. IBM challenged its research group to develop a system that could beat Jennings, or any other champion. It was the stuff of dreams for the researchers, who received in effect carte blanche to push the boundaries of AI.

IBM leadership saw an opportunity for financial gain, of course, and a public relations windfall (or tragedy). But management also seemed to have been bitten by the sci-fi bug, the futuristic idea of an "IBM-inside" language understanding system that beat humans to buzzers and rattled off all the right answers to questions, stumping most of the viewership at home. It was Deep Blue and chess again, but at the same time it wasn't. *Jeopardy!* is language-based. It's a question-answering (QA) system, technically, with the simplification that the question is put in a single sentence, and the answer always begins with "What is," and ends with the answer in a single phrase. This seems much more closely tied to human intelligence—it reminds us of the Turing test. *Jeopardy!* appeared to be an opportunity to leapfrog ahead in the AI arms race. (A fly on the wall might have gotten excited, overhearing management on that day in 2004 at IBM Research: "We're gonna make history. Money isn't an object. Find a way to build a system that plays *Jeopardy!*")

IBM assembled a team of top talent, and spent the next three years doing due diligence on the requirements for a question-answering system capable of playing world-class *Jeopardy!* This meant that it was open-domain QA, because the game covers diverse topics. (This is part of its appeal.) The system couldn't be hacked or hardcoded for specific topics, it seemed. They dubbed the future system Watson (after the founder and first CEO of IBM, not the fictional sidekick to Sherlock Holmes).

An older system called PIQUANT gave them a head start. It was built in 1999 by IBM to compete in the Text REtrieval Conferences sponsored by the National Institute of Standards and Technology. PIQUANT was a consistent top performer in the Text REtrieval Conferences competitions, but they were simplified question-answering games, and *Jeopardy!* was by contrast a broader and vastly more difficult challenge. For instance, PIQUANT answered questions with a predetermined set of labels, such as PERSON, PLACE, DATE, or NUMBER. Given a text supplied to the system as a question, PIQUANT would output the label representing the relevant topic. *Jeopardy!* play ranges over a vast number of topics.

PIQUANT could not be extended for this challenge, and there were other problems. In *Jeopardy!* a contestant should not "buzz in" with an answer unless he or she is very confident it is correct. Mistakes are penalized. Thus the type of QA required by computer *Jeopardy!* was importantly different. This meant, among other things, that the system would require a comprehensive redesign. The new system had to "grow" into the game of *Jeopardy!* by repeated trial and error, while PIQUANT was a one-size-fits-all QA system.

Dave Ferrucci, head of the Watson project at IBM, undertook the comprehensive redesign. Early on he acknowledged that playing *Jeopardy!*—still a game, keep in mind, and not open-ended reading—required a system not plagued too much by context in natural language: "Consider the expression 'That play was bad!' What sort of thing does 'play' refer to? A play on Broadway? A football play? Does 'bad' mean 'good' in this sentence? Clearly, more context is needed to interpret the intended meaning accurately."[8]

Ferrucci knew also about AI pioneer Marvin Minsky's ideas that complex intelligent problems are solved in polyglot fashion, by minds (or computers) with many submodules that break difficult problems down into manageable chunks. The task is then to combine the answers from the many modules into a "global" answer or solution. Minsky called this the "Society of Minds" approach to AI, and Ferrucci adopted it as the design inspiration behind Watson.[9]

The Watson team developed DeepQA, a QA system that generated many possible answers to questions and returned the best one based on multifarious analysis. DeepQA essentially implemented a software version of the society-of-minds idea for the *Jeopardy!* game. The system minimized too-quick answers to questions by pushing questions through a pipeline and holding off on scoring answers until all available evidence was collected. A piece of the puzzle for answering a question might be downstream in the DeepQA pipeline, for instance. This architecture of Watson was part of its eventual "intelligence." It is one reason that Watson, after years of development, became a significant hybrid solution to a complex, language-based problem.

There are other reasons. For one, an enormous amount of human analysis went into the design, development, and testing of Watson. This human contribution to successful system design is often overlooked, particularly when the application is supposed to be a showcase of AI. In fact it's clear that Watson was as much a product of careful and insightful game analysis by the engineering team as it was of improvements over PIQUANT or QA systems generally. Thousands of *Jeopardy!* games were analyzed. A baseline performance was established by first modifying PIQUANT to play *Jeopardy!* (It was, predictably, very bad. The PIQUANT system scored a depressing 16 percent correct on attempted answers to 70 percent of the questions posed to it, known as 16@70 by the IBM team.) The DeepQA module was then redesigned, targeting *Jeopardy!* questions to find specific clues indicating possible answers. The trial-and-error process of tailoring the DeepQA pipeline specifically to *Jeopardy!* was critical to eventual success. The IBM engineers even gave it a name, AdaptWatson.

AdaptWatson was not Watson—it was the human process of making Watson better, by zeroing in on *Jeopardy!*-specific tricks to include in DeepQA. In all, over one hundred special-purpose language processing modules were designed, deployed, and refined using AdaptWatson as a protocol. It was a massive workflow process that at its peak involved twenty-five researchers, including student help from local universities. From an engineering standpoint, AdaptWatson was brilliant: PIQUANT was retired, and an entire "*Jeopardy!*-optimal" pipeline took shape around game performance. DeepQA became an effective feedback loop involving dozens of expert humans gradually fine-tuning the Watson system to play world-class *Jeopardy!* (Perhaps less appreciated, the resulting system became narrow—a consequence itself of this engineering strategy aimed at game-play specifics.)

Inspired by the "society of minds" idea, DeepQA did include innovations in natural language processing (NLP). Older QA systems like PIQUANT relied on processing pipelines that were similar to IBM Watson: first analyze the question, then search and retrieve possible answers, then score the retrieved answers, returning the best.

Yet PIQUANT and other earlier-generation systems share a common flaw; they can get locked into incorrect answers by mistakes made in early parts of analysis. The Watson team modified an old architecture developed by IBM between 2001 and 2006 known as the Unstructured Information Management Architecture. UIMA is plug-and-play: software modules can be dropped into the pipeline, taken out and modified, then dropped back in. Likewise, entire algorithms can be swapped, so the pipeline speeds and facilitates trial-and-error runs and extensive testing necessary for large and complicated projects. Using UIMA, the Watson team built a more sophisticated pipeline for playing *Jeopardy!*, continually tuned by AdaptWatson. The hybrid system worked: Watson began playing world-class *Jeopardy!*

Yet, like its narrow predecessor Deep Blue, Watson was designed from the start to be good at one thing. The details of the Watson system are understandably complex, but even a brief pass through the development of the system and its key components dispels pretensions of real understanding. For instance, DeepQA relies on a grab bag of relatively well-understood techniques in NLP research, like parsing sentences, performing some types of co-reference resolution (resolving pronouns to antecedents, as we saw earlier), identifying named entities as person or place, and classifying questions themselves by types, such as the "factoid" type. The Watson team used these techniques but applied them specifically to *Jeopardy!* play. They also wrote a "QSections" module, which looks for obvious and game-specific constraints on answers in the questions themselves. For example, the phrase *This four-letter word means* . . . in a question signals that its answer will be a four-letter word. This is, again, design and development for optimizing game-play, rather than for general natural language understanding by a machine.

The magical-seeming quality of the Watson system is perhaps most quickly dispelled by looking more closely at the results of its search for answers. Lookup and retrieval of possible answers succeeds largely because of an exploitable shortcut in *Jeopardy!* discovered by engineers using AdaptWatson to identify problems and possible improvements: fully 95 percent of all answers to *Jeopardy!* questions are just Wikipedia titles. This serendipitous discovery made the entire effort possible; simply matching questions to Wikipedia titles would by itself result in superhuman-level gameplay. As usual, the devil is in the details.

The Watson system is impressive, but as with IBM's prior success with the chess-playing Deep Blue, it's unclear whether to assign praise to the supercomputing resources (Watson used over 200 eight-core computer servers) or to the insight and diligence of its human engineers—who are, after all, well-funded and able to work specifically on such a marketing windfall for IBM. To the team's credit, Watson was designed to perform more open-ended search on unstructured sources like blogs, digital bibles, and other sources. And Wikipedia itself is still mostly unstructured, although a database called "DB-Pedia" exists, which Watson included. But the search was, again, specific to computer representations of questions, and returned passages to be analyzed using techniques going back decades in research on language processing: filling in open slots in the question with words and phrases deemed a good fit by the results of the search. This is not nothing, of course, but it's diminished in significance, once again, by the realization that the most open-ended text analysis by Watson was also the most unreliable—another major demystifying observation. Lacking Wikipedia title matches, the capabilities of the Watson system were much less impressive (but then, these were only 5 percent of all answers). *Jeopardy!* itself has a trick, it turns out, because it's a factoid game, and factoids can be retrieved. This feature of the game, discovered using the human protocol AdaptWatson, accounts more than anything for its eventual superhuman performance—and also explains why IBM's foray into healthcare has been decidedly less triumphant.

Predictably, the success of Watson prompted discussion about the coming age of AI that truly understands natural language. But an ostensibly simpler task than playing *Jeopardy!,* like reading the newspaper (certainly, simpler for humans), had early on been ruled out by Ferrucci and his team. He said so explicitly, pointing out that even the contents of Wikipedia pages (not just the titles) posed problems for AI so difficult that general-purpose conversion of open-ended text into computer readable-form was intractable—a goal that was considered briefly, then abandoned.[10] The Watson team instead identified a set of high-value targets in information sources—excerpts and passages containing likely answers—that fit into the processing pipeline of Watson and raised the probability of correct answers. In taking this approach, the team effectively retired the game of *Jeopardy!,* like chess earlier. But it also proved once again that increasingly evident maxim, that all successful AI is narrow. (And it proved the corollary maxim, too: success at games generates public excitement, without advancing us toward artificial general intelligence.) Watson is not a step toward general intelligence, but rather further evidence that the quest for generality remains mired in mystery and confusion. While the IBM team did score an impressive victory using a powerful hybrid system, it did not discover a key to language understanding. All the problems of programming abductive inferences for general intelligence remain.

THE NARROWNESS TRAP AND LANGUAGE

I've focused on language understanding at some length (rather than, say, problems in robotics) because it so obviously reveals AI's narrowness trap, in which all known systems get caught. In what follows, I will review some recent applications that tout language understanding capabilities they don't have. They are, rather, examples of narrow AI, too often masquerading as more.

Take Google Duplex. Released in 2018, Duplex makes phone calls on behalf of its owner to accomplish routine tasks, like making reservations and appointments. Duplex has a human-sounding voice (which later prompted Google to identify the system as automated to callers, after public pressure mounted). Duplex seems like the coming of HAL from the movie *2001: A Space Odyssey,* until we learn that the system, developed with all the vast computational and data resources of Google, promised only to make restaurant reservations, book hair salon appointments, and find the opening hours of a few selected businesses. This sounds pretty narrow. It gets worse. After the demo, Duplex was released on Android phones without the option for open-ended reservations or inquiries about business hours. It only made reservations at restaurants. As Marcus and Davis put it, "It doesn't get narrower than that."[11]

Duplex is joined by a host of other recent offerings using big data and machine learning that promise HAL-like abilities but are caught in AI's narrowness trap, as well. Speech-driven virtual assistance like Siri, Cortana, Google Assistant, and Alexa answer questions posed by people, and can even engage in tidbits of banter, like giving humorous responses to playful (or insulting) questions. But their understanding of natural language is a facade, as anyone interacting with the systems knows. Like Watson, all are best at factoids that can be culled from databases and info-boxes (as with Wikipedia) on the web. *Who won the Super Bowl in 1975?* is a good question. But *Can a shark play checkers?* is not (recall Winograd schemas).[12] In general, questions that probe, however lightly, under the layer of facts pulled from the web, so that some real knowledge and understanding would be required, consistently flummox such systems. Their abilities begin and end in the paper-thin layer of collected facts and canned jokes on offer. Like Goostman, they have no real understanding, and so can't really connect with us—or, too often, help us.

Narrowness is endemic to systems like Watson that tackle natural language. As we saw in prior sections, this is because acts like reading and conversing are actually deep, open-ended feats of inference, requiring understanding of the world. Google Talk to Books, showcased predictably with much fanfare by Ray Kurzweil in a TED talk in 2018, promised an unparalleled question-answering capability by, as *Quartz* put it, "reading thousands of books."[13] In fact, it *indexed* about one hundred thousand books, encoding sentences numerically in vectors (data structures), and using deep learning (what else?) to compute their similarity to other vectors. This is a fancy version of the frequency assumption and the empirical constraint, all over again.

After the flashy TED demo, Talk to Books's many limitations quickly surfaced. Details and factoids about novels in its database were retrievable, but rarely questions requiring real inference, like abduction. If *The Great Gatsby* got indexed, for instance, it might return an answer for a query about the author (F. Scott Fitzgerald), or even about Gatsby's first name (Jay). But easy questions about plot and characters requiring knowledge-based inference quickly outstrip the capability of the system. Having read the novel, it's easy to answer the question "In what city did Gatsby first meet the protagonist of the novel?" For Talk to Books, answering such questions would demand inferential powers beyond its reach. Marcus and Davis asked Talk to Books *Where did Harry Potter meet Hermione Granger?* and received answers that weren't even in *Harry Potter and the Sorcerer's Stone,* and (even worse) omitted the central topic of the question—a location, specifying where the meeting occurred.[14]

Narrow performance is a problem for all systems tackling natural language understanding, not just Google Talk to Books. Language is about the world "out there," which involves necessary knowledge and a grasp of what things mean. As we've seen, too, the narrowness trap is a consequence of the data-driven approach itself, from its inception. To paraphrase the comic movie character Ace Ventura, "Narrowness is AI. AI is Narrowness."

Other examples are by now famous, or infamous. In 2016 Microsoft released its much-hyped chatbot, Tay. The software giant touted Tay as a quantum leap ahead of older, rule-based systems like the notoriously human-seeming ELIZA of 1960s yore, by actually learning from user interaction and data online. But the lessons of induction and its limits had not been learned, apparently, as Tay happily ingested racist and sexist clickstreams trolling it, along with other hate speech found on the web. Tay was quite the big-data student, blasting off tweets including "I fucking hate the feminists" and "Hitler was right: I hate the Jews," much to the dismay of Microsoft, who cancelled the hating Tay in less than a day.[15] But the outcome should have been foreseen, given the essential "garbage in, garbage out" nature of its chosen design. Tay was a case of corporate myopia about its own technical approach—and yet another example of narrow AI. In this case, genuine understanding would have bestowed upon Tay a minimal ability to filter out offensive tweets. But because it had no real understanding of language or "tweets" in the first place, it spit out what it took in. Tay is a memorable (but alas, forgettable) example of the idiot savant nature of data-centric AI.

Natural language understanding may be hard, but it's apparently irresistible, too. Facebook got in line for coming disappointment with the announcement of a system that could read "a synopsis of *Lord of the Rings* and answer questions about it," as *Technology Review* put it. But the synopsis was four lines of simple sentences like "Bilbo traveled to the cave. Gollum dropped the ring there. Bilbo took the ring." And the system could only answer questions directly answered in the sentences, like *Where is the ring?* and so on. Questions requiring an understanding of the text weren't possible. In general, questions probing answers to *why* questions weren't possible—say, *Why did Bilbo travel to the cave?* The system reduces *Lord of the Rings* to a few lines of text, a synopsis, and answers only the most mundane and stupid questions, lacking any understanding all the while. Narrowness is baked in.

The inference framework helps make sense of the narrowness trap. Watson, a system we looked at in some detail, turns out to achieve an impressive performance on a complicated game involving some clever treatment of unstructured information—notably web pages and especially Wikipedia pages (and, here, especially titles). A deeper dive into the system reveals its hybrid design, from hard-coded rules to statistical methods for scoring answers, to Monte Carlo methods for betting in "Daily Doubles" and Final Jeopardy (betting was left out of the discussion above). Inferentially, question analysis in Watson relied mostly on traditional techniques for tagging sentences with parse, entity, and other information—rules, in other words, or deductively-inspired techniques. QSections, like *This four-letter word means . . .*, which are typical of *Jeopardy!* questions, are easy to handle without statistics. Similar remarks apply to other aspects of question analysis that can be reliably identified by inspection of questions. Why not use machine learning? Because many problems that are quite simple for rule-based or deductive-logical approaches pose intractable problems for machine learning. Thus Watson was a clever hybrid. And yet, demonstrably, it is caught in the narrowness trap anyway. The overarching explanation of the trap is simply that general inference, not available, can't be made up for using combinations of rule- or learning-based approaches. To put it another way, lacking abductive inference, system performance must be narrow—general intelligence is not available. Narrowness is inevitable.

For what it's worth, Watson employed an impressive set of machine-learning techniques. Some 25,000 *Jeopardy!* questions were analyzed, converted into 5.7 million training examples for the system. The system produced question-answer pairs, accumulating evidence in the processing pipeline, and scored the list of pairs using statistical techniques—all possible because of data about past games, where the outcome is known.

It's notable that Watson didn't use deep learning—at least not the version of Watson that outplayed human champions in the televised event in 2011. Deep learning would not have helped—and here again, this is a nod to the ingenuity of the IBM team. A relatively simple machine-learning technique known as regularized logistic regression was used, even though more powerful learning algorithms were available. (Deep learning in 2011 was still relatively unknown.) More powerful learning systems would simply incur more computational training and testing expense—AI is a toolkit, in the end. The Watson system had no real innovation in any particular technique, but in combining them in the "society of minds"-inspired framework of a decomposable pipeline (using UIMA), it achieved world-class results. Inferentially, Watson is perhaps the best example to date of the power of using all available deductive and inductive approaches in AI, combined in a smart architecture. But, take out the trick of Wikipedia titles, and it would not have succeeded. It's still narrow—very narrow—in the end, like all other known systems, hybrid or otherwise.

We might coin another term to explain all this: call it the inference trap. Since the three known types of inference are not reducible to each other but are distinct, and abductive inference is required for general intelligence, purely inductively inspired techniques like machine learning remain inadequate, no matter how fast computers get, and hybrid systems like Watson fall short of general understanding as well. In open-ended scenarios requiring knowledge about the world like language understanding, abduction is central and irreplaceable. Because of this, attempts at combining deductive and inductive strategies are always doomed to fail—it might just take longer to figure out why, as in the case of Watson. The field requires a fundamental theory of abduction. In the meantime, we are stuck in traps.

LOGICAL PIANOS AND VOYAGES

Charles Sanders Peirce knew about the possibility of using machines to explore logical inference. Computers didn't yet exist, but ideas about them did, and some physical devices had been built. The British logician and philosopher John Venn (the creator of the eponymous Venn diagrams) had speculated about building a purely automatic logic machine. And one of Peirce's students at Johns Hopkins University, Alan Marquand, had in fact begun working on a logical machine in 1881. Marquand had begun extending, in effect, a proto-computer, known as Jevons Logical Piano, after its inventor, the Englishman William Jevons. Marquand's machine was intended to solve problems in deductive logic, an area that Peirce had spent much of his life studying. Peirce himself took an active interest in the development of the logic machine, sketching out designs for the electromagnetic operation of Marquand's contraption.

Writing about the experience in 1887 in an oddly prescient paper titled "Logical Machines," in the *American Journal of Psychology*, Peirce begins characteristically with cautionary comment. "In the 'Voyage to Laputa' there is a description of a machine for evolving science automatically," he writes. "The intention is to ridicule the Organon of Aristotle and the Organon of Bacon by showing the absurdity of supposing any 'instrument' can do the work of the mind." Peirce, the skeptic, no doubt appreciated the wisdom of Swift's imagination. But he was sufficiently taken with the Promethean spirit to highlight the important work he and Marquand undertook. He credited his pupil, and his pupil's predecessor: "Yet the logical machines of Jevon and Marquand are mills into which the premises are fed and which turn out the conclusions by the revolution of a crank." The American inventor Charles Henry Webb, too, had designed a machine for performing arithmetic, and the English genius Charles Babbage developed a proof of concept (along with his protégé Ada

Lovelace) for a more visionary machine, performing general computations. They were machines that could "perform reasoning of no simple kind."[16]

"Logical Machines" then departs into detailed discussion of automating deductive syllogisms. At the end of the paper, tying up loose ends, Peirce comments on the possibility of what we now call Artificial Intelligence. "Every reasoning machine, that is to say, every machine, has two inherent impotencies. In the first place, it is destitute of all originality, of all initiative. It cannot find its own problems; it cannot feed itself. It cannot direct itself between different possible procedures."[17] Peirce then cites a complicated logical problem whose solution requires the selection of premises through dozens of steps. Perhaps the example can be solved somehow, some way (it probably can be today). Peirce allows for this possibility, but it doesn't matter. It misses the point. "And even if we did succeed in doing so, it would still remain true that the machine would be utterly devoid of original initiative, and would only do the special kind of thing it had been calculated to do."[18] Like much of his thinking, Peirce here joined a discussion that began in earnest decades after his death.

He adds, too, a simple idea that seems still mired in confusion, coloring science with mythology. "This, however, is no defect in a machine; we do not want it to do its own business, but ours." The trap of narrowness, too, was to Peirce an obvious feature of machines: "the capacity of a machine has absolute limitations; it has been contrived to do a certain thing, and it can do nothing else."[19] The scientist who spent his life's work exploring the mystery of human intelligence knew all too well that machines were, by design, poor and unsuited replacements. Swift's fantasies held wisdom.

In the next century, Turing proposed that we take up the challenge of infusing machines with "original initiative," by first programming them to talk to us. Turing was aware of Peirce's objection, which he attributed to Lady Lovelace in his 1950 paper. He also had played with simple learning algorithms, and in the decade of the 1950s single-layer neural networks appeared (called a perceptron). Understandably, Turing thought perhaps we could escape Peirce's and Lovelace's objections by creating learning machines modeled on the human brain. Reading "Computing Machinery and Intelligence," one gets the impression that learning represented the only real escape from the inherent limitations of machines, and the only real hope for passing the Turing test.

It didn't—it hasn't happened. Believing that it will, that it must, has consequences for society that now have become all too apparent. In this book's final part, we look at some of the consequences of the inevitability myth—particularly its deleterious effect on science itself.

Part III

THE FUTURE OF THE MYTH

Chapter 15

...

MYTHS AND HEROES

Ideas have consequences. In the following chapters, I hope to convince you that the consequences of the myth of artificial intelligence pose a significant and even grave threat to the future of scientific discovery and innovation—and, ironically, to progress in the field of AI itself. This final section is about our future, but we must begin in the past, for the problem of creating life, of designing artificial intelligence—literally, a mind in a machine—has always been infused with a sense of mythology, of humans reaching beyond themselves and attaining godlike power. The myth of AI is Promethean.

THE PROMETHEAN MYTH

Prometheus stole fire, which represents life, from Zeus and used it to cook meat for everyone on Earth. No surprise, Zeus was angered and came down to Earth to demand his fair share of every animal that mankind cooked. Prometheus then tricked Zeus into choosing as his portion only the entrails and guts.

Zeus did what gods always do when someone attempts to usurp their power and authority—he punished Prometheus, binding him to a rocky cliff and sending an eagle (the emblem of Zeus) to eat his liver. Every night, Prometheus's liver grew back; every day, the eagle returned to eat it again.

The story is about the expansion of human powers. It's a testament to the inherent and seemingly inexhaustible creative spirit in human beings. It's also a story about hubris. Prometheus might have kept his fire and spared his liver by offering Zeus up the best cuts. Our deep longings for true AI draw inspiration from the Promethean myth. We want to steal fire from the gods, despite potentially horrific consequences—eternal punishment, no less.

Prometheus was a punished hero, which is why Mary Shelley titled her enduring novel *Frankenstein: Or, a Modern Prometheus.* Frankenstein has been Hollywoodized over the decades and transformed into a farcical tale of a green monster. But it's really a story about the Promethean spirit in human beings and its consequences. A young and recently married Mary Shelley conceived of Frankenstein in a nightmare in a hotel in Switzerland, after late-night discussions with her husband, poet Percy Shelley, and Lord Byron, another poet. Her dream images never provided a blueprint for creating the monster, so from the beginning her creation asked the human question at the heart of the modern myth: What if such a being were possible?

Dr. Frankenstein, through an unspecified congeries of occult methods such as galvanizing dead tissue, performs the Promethean miracle: he creates intelligent life from dead matter. The story is a progenitor of later and specifically mechanistic portrayals of the creation of intelligence using science and technology, and it's also—and importantly—a deeply human story about spiritual isolation. Dr. Frankenstein is a mad scientist who possesses hidden and forbidden knowledge, which lets him play God. His creature comes alive, with consciousness and a longing for a romantic partner. Inevitably, Dr. Frankenstein's world falls apart, just as Prometheus's does. Mary Shelley, writing at age nineteen, captured the ancient myth and brought it alive into the modern world, as did Percy Shelley when he later wrote his famous *Prometheus Unbound,* a story of liberation. The Romantic writers give effective voice to the endless struggles and pains of the human condition, which is why we still talk about their creations and see, in our science and our scientists, both their dreams and their warnings.

Turing was not a "mad scientist" in Shelley's sense. Bletchley was a collaboration, and Turing's later efforts at building the world's first universal electronic computer were also team-based. He witnessed in Von Neumann, in America, a talent that would ultimately beat him to the punch, armed as Von Neumann was with rare genius and the deep pool of scientific and financial resources at his disposal. Turing—and Von Neumann—were scientific adventurers, but they worked within an environment that supplanted and suffused their genius with other talents.

Still, Turing probably had something like the Promethean idea swirling about as he thought about AI, which (recall) he believed might engage in genuine conversations by the year 2000. He knew enough Promethean geniuses in his own day, men who reached beyond mankind's ordinary grasp: Einstein, the brilliant logician Kurt Gödel, and of course Von Neumann. Once the computer formalism—Turing's eponymous Machine—was "out there" for science, then some scientist, perhaps working in a Bletchley environment, might crack the secrets of the human mind and write them up in code. For scientists don't believe in vagaries like the "evolution of science" except as frosting, as backdrop. They really believe in scientific genius. They really are all possessed by Prometheus, by what innovators can dream and achieve.

As work on AI keeps hitting hurdle after hurdle, however, the Promethean myth of astonishing innovation by individuals is disappearing from cultural archetypes in research and the broader culture. In its place, we have a passive-evolution mythology about AI that grows as belief in human potential shrinks.

Thoughtful critics like Jaron Lanier give voice to the central problem, "We should seek instead to inspire the phenomenon of human intelligence."[1] But, already, there are no more heroes. Instead, we have "hives."

OF HIVES AND MACHINES

Shifting the locus of intelligence from humans to machines is a gambit—a chess term, meaning the sacrifice of material for better position—which unavoidably has consequences for human culture. We would be forced to accept this gambit if the scientific and empirical evidence for it was unavoidable—suppose superintelligent aliens arrived, and quickly outsmarted everyone and took over. Absent such evidence, it's a ploy that leads to a diminished culture of innovation and progress. Why sacrifice our belief in human innovation, if we don't have to?

The ploy is, ironically, conservative; when smartphones are seen as evolving into superintelligence, then radical invention becomes unnecessary. We keep in place designs and ideas that benefit the status quo, all the while talking of unbridled "progress." Human intelligence becomes collective, like a hive of bees, or worse, the hive mind of Star Trek's Borg Collective, always organized by some invisible someone behind the scenes. Basically, in this mythology the human mind becomes an outdated version of coming machines.

But as we've seen, we have no scientific reason to believe any of this, so we shouldn't play mythological games with real life. We should build technology to push ahead a frontier of our own choosing.

In the first decade of the new century, we thought that's what we were doing.

THE RISE OF MACHINES (WAS: THE RISE OF PEOPLE)

When "Web 2.0" burst on the scene with a host of new technologies for "user generated content" like wikis and blogs, many culture and technology critics assumed we were in the midst of an explosion of human potential, a new era of possibility. In 2005, AI was still nursing wounds from the last "Dot Com" bubble of 2000, and machine learning and big data weren't yet ready for hype-mode. Citizen Bloggers were. An entire school of thought emerged, circa 2005, viewing the web and specifically Web 2.0 technologies as the new printing press, destined to liberate the intelligence and creativity of humans. The web promised not only to make us smarter and more informed, but enable us to collaborate more effectively, building modern-day digital pyramids and transforming science and culture. As I write this in 2020, however, the original Web 2.0 ideas have already disappeared. In fact, they seem downright surreal.

Clay Shirky, a writer and consultant who is now a professor at New York University's Interactive Telecommunications Program, once penned Web 2.0 era best sellers like *Here Comes Everyone* and *Cognitive Surplus: Creativity and Generosity in a Connected Age,* portending the rise of an uber-informed, socially conscious citizen, a new persona.[2] A little jingoistic, his message was still clarion: web denizens were poised to rewrite the rule books, ridding the world of stodgy "gatekeepers" like mainstream press and media, who unfairly controlled the production and flow of news and knowledge. "Power to the people" was the trope of the mid-2000s, a meme that copied and spread itself endlessly in blogs, commentary, and on bookshelves (and in e-books).

Yochai Benkler, Harvard University professor of Entrepreneurial Legal Studies, proclaimed in his widely read *The Wealth of Networks: How Social Production Transforms Markets and Freedom* in 2006 that a new era was upon us, a kind of revolution where large numbers of networked people would take on collaborative projects online, all for the public good, without requirements like paychecks.[3] Wikipedia seemed to buttress his point, a case of collaborative production without expectation of financial recompense. *Wired* editor Kevin Kelly (and others) later called Benkler's paean to online collaboration a hive mind, a nod to the social intelligence of bees, without a whisper of irony or derision. Benkler himself prefaced his academically serious rallying cry to the Web 2.0 world with a quote from John Stuart Mill: "Human nature is not a machine to be built after a model, and set to do exactly the work prescribed for it, but a tree, which requires to grow and develop itself on all sides, according to the tendency of the inward forces which make it a living thing."[4]

It's an excellent quote. But Mill's words have an outlandish pie-in-the-sky feel to them today, in large part because his center of gravity is the human person rather than a machine.

Shirky's ideas, too, have a whimsical and naive feel to them now. Cognitive surplus captures the insight that, when everyone goes online, they might quit or cut back on mind-numbing activities like watching sitcoms. There's a surplus of cognitive—thinking—power in the age of the internet, which we can turn to good use, like bringing about social revolution in an Arab Spring, or inventing cures for cancer. Shirky's precursor book, *Here Comes Everybody,* bustled with anecdotes about ordinary people helping the police capture crooks by using mobile technology.[5]

We can, of course, still pitch in like this with smart phones, but the Venn diagram of everyday usage no longer points to a future of realized human potential. In fact, it's clear that the intellectual revolution prophesied in the mid-2000s never happened. (The "hive mind" didn't even give us Wikipedia—most of the real writing is done by singular experts, with others performing more mundane editing tasks.) Shirky's and others' optimism about human growth transmogrified rather quickly into a worldview that sees humans as cogs in a giant machine. Eventually, the machine itself—the network, the system—becomes the focus. Predictably, hive minds ended up promoting a new skepticism about human intelligence. The idea now feeds popular mythology about the ascendancy of machines. Supercomputers have become "giant brains."

If we had to pick a year that "human potential" died, as a serious online meme anyway, 2008 would be the frontrunner. "Big data" itself entered the lexicon. Chris Anderson of *Wired* published his provocation that big data would replace theory in science—a not-so-subtle suggestion that human innovation could simply be outsourced to computation.

And AI by 2008 had been repackaged in its modern guise as data science. The trajectory here in retrospect seems obvious: from citizen bloggers—individuals—forging a new human future; to hive minds buzzing about, making encyclopedias; to big data and AI replacing human thinking, even ridding us of pesky theory in science. "Human nature is not a machine," said Benkler quoting Mill just two years earlier. Stunningly, Benkler's hope for humanity online got subsumed into mythology about AI—a machine revolution—that now displaces and ignores human creativity. Like much discussion about AI these days, the transformation seems foolishly motivated and conceived.

Scientists and other members of the intelligentsia eventually pointed out that science without theory doesn't make sense, since theoretical "models" or frameworks precede big data analysis and give machine learning something specific to do, to analyze. But the zeitgeist of mid-2000's Web 2.0 had turned abruptly away from "power to the people" by 2010.

Two years later, in 2012, Deep Learning systems blew away competitors on the well-known ImageNet competitions using Flickr photo datasets, and quickly showed fantastic promise on other consumer problems like voice recognition and content personalization—problems that companies like Facebook (which that year went public to the tune of over sixteen billion) and Google needed to solve to sell ads and recommend content for their legion of users. Facebook, Amazon, Google and the other tech giants quickly embraced big data AI, and soon everyone forgot about citizen bloggers. The intelligentsia began extolling a coming AI that would blog and write news for us. Next, they would replace us. In retrospect, Lanier's worry in his 2010 *You Are Not a Gadget* was prescient, but all too late: "A new generation has come of age with a reduced expectation of what a person can be, and of who each person might become."[6]

Perhaps the hive mind idea itself seems a little quaint today, if only for the equally depressing reason that major ideas about human potential have receded. In 2005 Google was still a marvel, a wonderful example of human innovation. Today, our ubiquitous search engine giant is like the keychain in our pocket. We have ceased to even notice it. Less than a decade after James Surowiecki's 2005 hit *The Wisdom of Crowds,* the idea that people on Twitter or other social networks display collective wisdom—or wisdom at all—seemed laughable.[7]

It is telling that mythology about AI has not also been ridiculed, and seems ever more on the rise.

Chapter 16

...

AI MYTHOLOGY INVADES NEUROSCIENCE

"Hive mind" remains in the lexicon, even if it isn't invoked with such seriousness anymore. But its offshoot has appeared in the most unlikely of places, in science itself. Hives for minds, then "swarms" for scientific discovery.

Take Sean Hill, formerly the director of the International Neuroinformatics Coordinating Facility, part of a major collaborative effort known as the Human Brain Project. Writing for the 2015 anthology *The Future of the Brain,* Hill sees large-scale collaborative efforts as the future of science, and individual scientists as best understood as part of swarms: "One goal of the Human Brain Project is to trigger and facilitate a new wave of global collaboration in neuroscience. . . . If successful in engaging the community, the aim is to have swarms of scientists attacking the major challenges of understanding the brain and its disorders together—in an environment where every individual will receive credit for his or her contribution."[1]

This is a hodgepodge of ideas, from "global collaboration," which sounds promising, to "swarms of scientists," which evokes an absurdly deflationary metaphor for individual scientists' contributions (effectively disallowing individual discovery itself), to a tack-on bromide about "every individual . . . receiving credit."

Perhaps it was an off day for Hill, a major player in the now infamous Human Brain Project underway in Europe. But Henry Markram, once director of that project, is also a proponent of Hill's vision of science, arguing that geniuses like Albert Einstein are now unnecessary: "We are hampered by the general belief that we need an Einstein to explain how the brain works. What we actually need is to set aside our egos and create a new kind of collective neuroscience."[2] But his touting of collective neuroscience was, we now know, his own mythological vision of creating a superintelligent computer brain, using other scientists as resources to pursue a definite but ill-advised path.

HUMANS NEED NOT APPLY

Rhetoric about swarm science, like talk of hive minds, inevitably leads to a computer-centric view of the world, where human potential is downplayed in favor of the ascendance of machines. Science is following online culture, from human ideas to mega technology, which has led to the consolidation of power in major tech companies and a general stagnation in the pace of innovation.

Web 2.0 futurists know this all too well. Platforms for "user-generated content" sparked, first, new visions of human possibility. Then, as the technology matured, came visions of connecting people in huge collaborative efforts, and finally the AI inevitability mythology, which relegates people to the sidelines in its narratives about the future of machines.

The same trend appears now in basic research. Science, once a triumph of human intelligence, now seems headed into a morass of rhetoric about the power of big data and new computational methods, where the scientists' role is now as a technician, essentially testing existing theories on IBM Blue Gene supercomputers.

But computers don't have insights. People do. And collaborative efforts are only effective when individuals are valued. Someone has to have an idea. Turing at Bletchley knew—or learned—this, but the lessons have been lost in the decades since. Technology, or rather AI technology, is now pulling "us" into "it." Stupefyingly, we now disparage Einstein to make room for talking-up machinery.

Large-scale neuroscience projects are an unfortunate case in point.

THE HUMAN BRAIN PROJECT

The Human Brain Project launched officially in October 2013 with a ten-year, $1.3 billion award from the European Union, a massive sum for exploratory neuroscience research. The project initially involved more than 150 institutions around the world, led by Dr. Henry Markram, a neuroscientist at the Swiss Federal Institute of Technology in Lausanne. Markram is known for his Blue Brain project, an ambitious attempt to model an entire neocortical column in a rat's brain in silica, in a computer simulation on an IBM Blue Gene supercomputer.

The Human Brain Project's goals expand Blue Brain's scope to include no less than a complete computer simulation of the entire human brain, a goal that Markram announced in a 2009 TED talk would be met by the end of the decade—though many other neuroscientists disagreed. Like futuristic claims made about AI, Markram's prognostications were wrong—very wrong—and fortunately for science, the failure of his predictions was not altogether ignored. Writing in *The Atlantic* in 2019, Ed Yong remarked succinctly on what other neuroscientists had been predicting all along: "It's been ten years. He did not succeed."[3]

Confronted with mismanagement allegations, Dr. Markram stepped down as head of the project two years after it launched. A few years later, the project rebranded itself, shamelessly, as simply a software project, providing tools and methods for human scientists conducting ongoing and potentially important research.

The swarm science idea for doing fundamental scientific research is a blatant error, and the now familiar move—conscious or unconscious—to diminish human involvement and refocus efforts on supercomputers and big data, extrapolating from existing technology, should be exposed, just as similar hype about AI should be confronted.

Troubling, too, is that Markram and Hill's ideas about the future of neuroscience are public examples of a worldwide trend of scientists attempting to push science forward with computation rather than with ideas. Markram and the Human Brain Project are perhaps the most egregious examples of a mythology about intelligence "emerging" from Big Science, whose centerpieces are supercomputers; but other projects, less publicly, make the same mistakes.

For instance, on the heels of the Human Brain Project award, the Obama administration announced an equally ambitious Big Science effort, the Brain Research through Advancing Innovative Neurotechnologies (BRAIN) initiative, investing an initial $100 million for the 2014 fiscal year with expected expenses of fully $300 million over ten years. The BRAIN initiative focuses on developing technologies that can model neuron circuits and other functional areas of the brain comprising multiple individual neurons. And smaller, yet significant, brain simulation projects like the Allen Brain Atlas from the Allen Institute for Brain Science in Seattle are also underway.

Such projects offer the promise of a complete understanding of the brain. Markram, for one, has said publicly for years that he plans to embody his supercomputer simulation in robotics and thus create the world's first non-biological intelligence. Along the way, the Big Brain projects please government agencies by promising more practical benefits, like insight into the causes of Alzheimer's and other brain-related diseases. President George H. W. Bush once declared 1990–2000 the "decade of the brain." It appears, rather, that our current decade has become just that. The question is whether any substantive, fundamental progress has occurred.

While the Human Brain Project and BRAIN initiatives are clearly Big Science projects—high-profile objectives, top-down management, well-funded, and with an engineering rather than theory focus—they are billed as AI projects, as well. In particular, large neuroscience efforts today are almost universally promoted as big data projects. The data requirements for these projects certainly merits the term. The Kavli Foundation notes that the BRAIN Initiative must survive a "data deluge": "Measuring just a fraction of the neurons in the brain of a single mouse could generate nearly as much data as the 17 mile-long Large Hadron Collider or the most advanced astronomical observatories."[4] Kavli highlights a theme that runs prominently in the literature of both Big Brain projects: the marriage of data-driven AI to neuroscience is both an information technology challenge and a huge opportunity, as the ability to manipulate more data about the brain is thought to translate to successful research.

Indeed, big data is a centerpiece of discussion about the Big Brain projects now underway. Markram himself, for instance, insists that the Human Brain Project is about data integration. (Not neuroscience?) And Amye Kenall writes in *BioMed Central,* discussing the Human Brain Project's sought after "new supercomputer," that "the neurosciences will easily far exceed genomics as a data-intensive science." As Kenall notes, current supercomputers run on at "peta-scale," yet the Human Brain Project is expected to require "exa-scale" computing resources, so the project, along with neuroscience research, is also funding the development of the first exa-scale supercomputer.[5]

This blurring of the line between computing and information technology and neuroscience research is typical of both projects. Given the stated goals of the Big Brain initiatives, the focus on artificial intelligence concepts and techniques is, of course, necessary. Both projects confront what is known as the brain-mapping problem, a problem that is inherently computationally complex. Rebecca

Golden, writing for the Genetic Literacy Project, explained the mapping problem this way:

> The human brain is estimated to have approximately 86 billion neurons (8.6 x 1010), each neuron with possibly tens of thousands of synaptic connections; these little conversation sites are where neurons exchange information. In total, there are likely to be more than a hundred trillion neuronal synapses—so a computer recording a simple binary piece of information about synapses, such as whether it fired in a time window or not, would require 100 terabytes. The amount of storage needed to store even this very simple information every second over the course of one day for one person would more than 100,000 terabytes, or 100 petabytes. Supercomputers these days hold about 10 petabytes. And this quick calculation doesn't account for the changes in connectivity and positioning of these synapses occurring over time. Counting how these connections change just after a good night's sleep or a class in mathematics amounts to a whopping figure (and many more bytes than the estimated 1080 atoms in the universe). The wiring problem seems intractable in its magnitude.[6]

Markram and other researchers are, of course, aware of the seeming intractability of the mapping problem in neuroscience, and herein lies his and others' continuing allegiance to Big Data AI as a prime mover. If new insights are to emerge from "data integration," as he puts it, then principles linking neurons together into circuits and larger functional units (mesa circuits) will constrain the mapping problem and simplify the computational complexity researchers now face.

In other words, big data and AI will supply the missing pieces of theory in neuroscience itself. Under this view, information technology, thought to be a distraction by critics of Big Brain projects, is actually part of the argument for why such projects will (eventually) succeed: the technology, AI itself, will fill in missing pieces, where humans have so far failed.

BIG DATA, AGAIN

The Data Brain projects, and in particular the Human Brain Project, were bold attempts at advancing a big-picture vision of science, tackling major questions about the nature of human thought. Indeed Markram, undeterred by the project's failure, unabashedly promulgates Data Brain projects as a route to AI. He believes that AI and neuroscience will crack the mystery of human intelligence, and perhaps our consciousness, too. In a host of public interviews over the years, Markram has stated that he plans to model neurons using data about the brain contributed from projects all over the world, to discover the "statistical principles" undergirding neuronal activity (at the level of ion exchanges), and to link larger and larger functional units of neurons in the human brain together until a complete map emerges.

This map, he thinks, will then be capable of exhibiting human-like behaviors. Futurists such as Ray Kurzweil and a bevy of other AI mythmakers persist in this belief, as well. We will, in other words, understand the principles of intelligent thought to the point where they can be reduced to engineering, programmed into robotics and AI systems. A new era of AI will launch on the heels of major advances in neuroscience. And, again, we will succeed where previous generations have failed because of AI: our access to volumes of data and data integration and analysis platforms that enable us to discover principles and theories, where before we were awash in disparate research.

AI, in other words, is what makes the Big Brain initiatives seem new and different from what came before—not breakthroughs in neuroscience theory, but computational capacity and data. But this is pure folly, as the failure of the Human Brain Project clearly demonstrates. Like induction compared to abduction, technology is downstream of theory, and information technology especially is filling in as a replacement for innovation, while researchers tout its breakthrough potential, implementing large tech frameworks using existing ideas.

It is probably true—or at least it's reasonable to assume—that inadequate neuroscience knowledge is one of the key reasons we don't yet have better theories about the nature of our minds. In particular, a better understanding of the principles of human cognition could inform AI, dedicated to unlocking the mystery of intelligence—a stated goal of Markram. Yet large-scale brain simulations (Data Brain approaches) would seem to get things exactly backward in our quest for such knowledge. Lacking a theory about how the brain behaves—how we think and feel and perceive—existing knowledge, about neurons and functional units like circuits expressed in computer simulations, pins hopes that the missing ingredients of cognition will somehow emerge from the volumes of collected data at these lower levels.

This is, of course, a core conceit in mythology about AI itself: that insights, theories, and hypotheses unknown and even unknowable at smaller scales will somehow pop into existence once we collect and analyze enough data, using machine learning and other inductive approaches.

The Data Brain efforts of the Human Brain Project and BRAIN initiative in the United States both endorse this dubious idea, hoping that neuroscience generally will follow the example of the Human Genome Project, proving that science can be reduced to engineering (and scientists to swarms of helpers).

But it should be noted that the Human Genome Project had well-defined goals which omitted major theoretical challenges—it was an engineering project from the get-go. This is not the case, fortunately, for all Data Brain projects. Researchers in neuroscience—from Eric Kandel at Columbia University to Christof Koch, formerly at Cal-Tech and now at the Allen Institute for Brain Science, to even Markram—all will admit that major pieces of theory are missing from our picture of the brain, particularly at higher levels of functionality. The response, coming most forcefully from Markram and the still-ambitious Human Brain Project, is that big data will fill in missing theory and that scientists are wasting their time performing research in the traditional manner—in small, well-defined research objectives on specific problem areas. In the age of AI, apparently, we can't wait for theory to come from discovery and experiment. We have to place our faith in the supremacy of computational over human intelligence—astoundingly, even in the face of an ongoing theoretical mystery about how to imbue computers with flexible intelligence in the first place.

This thinking is a mistake, and when the dust settles, will likely prove a costly one at that.

In spite of the constant rhetoric about progress, in fact, Data Brain projects are actually conservative in nature when it comes to neuroscience research itself. Researchers at the Human Brain Project, for instance, seem content to use existing neuroscience research as the basis for "data integration" plans, mistakenly believing that the data itself will provide answers as the complexity of the simulation grows—an original conceit of big data. Hence, the project and other like-minded AI-driven projects effectively undermine the process of scientific discovery by placing the emphasis on larger and larger simulations of existing experimental knowledge.

That high-profile Data Brain projects like Europe's Human Brain Project and now the BRAIN Initiative in the United States were able to convince funding agencies and, to a large extent, the unknowing public that such existing knowledge, simulated on supercomputing platforms crunching large volumes of data, constitutes a scientific advance rather than a conservative engineering project speaks to the general confusion about the role of AI in science generally.

For instance, a technique originally championed by Markram and adopted by the Human Brain Project is now, in various guises, widespread in major neuroscientific efforts generally. Called "predictive neuroscience," it's an approach by researchers to simulate connections between neurons. Unknown synaptic links are determined from known links using inductive machine-learning techniques. Both traditional neural networks and more powerful deep learning networks are used. Markram showed initially that machine learning can correctly predict previously unknown connections in the cortical columns of the rat brain. An analysis using a standard F-measure statistic yielded an accuracy of nearly 80 percent for the approach.

While representing an advance in applying machine learning to biological datasets, the approach has an average error rate of two in ten. This has ominous implications for any strategy for reverse-engineering the human brain. But this concern aside, it is the shallow logic behind such approaches that spell deeper trouble for the institution of science.

HIGGS BOSON

In 2012, scientists discovered a long-missing piece of the standard model of physics, the Higgs boson. The discovery of Higgs boson is often attributed to an impressive piece of technology, the massive Large Hadron Collider (LHC) straddling the French-Swiss border. The LHC is seventeen miles of tubing constituting the world's largest supercollider.

Scientists used the LHC to design an experiment specifically testing for the existence of a particle accounting for mass in the universe, dubbed the "Higgs boson" after Peter Higgs, the scientist who first predicted its existence. The result of the high-energy experiment using the LHC seemed to confirm Higgs's theory about the existence of the particle (it was officially confirmed in 2013).

But the case of the Higgs boson also illustrates the perverse tendency to downplay theory and champion computational methods that buttress wishful thinking about big data and AI.

It's true that supercomputing resources are necessary to make sense of the mountain of data generated by the supercollider. In 2012, the collider generated about twenty-five petabytes a year; by 2018 the number had doubled, equivalent to about fifteen million high-definition movies. There is no doubt that big data analytics and the computing resources necessary for processing such volumes of data give physicists a set of powerful tools for exploring the subatomic world. But the story of the Higgs boson, often touted as an example of Big Science success (and it was), is also a triumph of theoretical insight. Higgs's case is particularly impressive as a case for theory, not for big data per se. Peter Higgs actually discovered the particle in 1964; the LHC then confirmed its existence. This is a case study in the proper use of technology to supplant human insight. The lesson of the Higgs particle confirmation is not a call-to-arms for mythology in our computer future. Rather, it is a reminder that AI—here, big data—works only when we have prior theory. A great deal of confusion and potential trouble for science lies precisely with this point.

Unfortunately, neuroscience, unlike particle physics, has no unifying theoretical framework. The challenge, then, for neuroscientists is to defend the current data-driven model in the absence of theoretical insights making sense of and guiding the "deluge of data," as the Kavli Foundation puts it.

NEEDED: THEORISTS

In a revealing interview with *Nature* a few years ago, prominent neuroscientists Eric R. Kandel (Director at the Kavli Institute for Brain

Science at Columbia University), Henry Markram (Director of the Human Brain Project), Paul M. Mathews (Head of Division of Brain Sciences at Imperial College London, UK), Rafael Yuste (Professor of Biological and Neurosciences at Columbia University), and Christof Koch (Chief Scientific Officer at the Allen Institute for Brain Science in Seattle) discussed the role of big data, theory, and collaboration in Europe's Human Brain Project and the BRAIN Initiative in the United States.[7] It is clear in this discussion that neuroscience faces major challenges, and that answers are still in short supply. Mathews, for instance, admits that "the BRAIN Initiative and the Human Brain Project both face a fundamental challenge: we do not have a strong paradigm to guide inquiry. It is striking that both the BRAIN Initiative and Human Brain Project are 'big data' collection exercises from which meaningful relationships are anticipated to emerge."[8]

Yet Markram remains confident that Big Data AI will fill in missing pieces of theory as the efforts progress. He is explicit about this, claiming that "the more data we have, the more biologically accurate the models will become." He makes clear in the *Nature* interview his vision of the Human Brain Project and of neuroscience generally: "Scientifically, we want to open the road to a new form of accelerated neuroscience in which we identify basic principles spanning multiple levels of brain organization and exploit these principles to fill the large gaps in our knowledge. For instance, we can use principles about the way neurons connect to predict the connectome [the wiring diagram of the brain]. Hypothetical reconstructions of the brain can guide and accelerate experimental mapping of the brain, turning it from a dream into a practical reality."[9]

It is clear from reading Markram's voluminous press interviews that he believes data integration—relevant research results from around the world collected into his supercomputer-based technical platform—will facilitate a "theory emergence" at higher and higher levels of brain organization. Hence, the initial neuron models using information about ion channels (that is, low-level molecular information about individual neuron behavior) will help drive theories of how functional units like neuronal circuits operate, which will in turn suggest principles or theories enabling researchers to connect microcircuits to mesa circuits to whole brain regions like the neocortex.

Markram himself is very clear about how this will work, and he's equally dismissive of suggestions that creative research by smaller research teams might find such theories independent of his Big Data approach. His interests are in using scientists and their research as inputs into a centralized technology framework, a motivation he openly admits, believing it will usher in "a new kind of collective neuroscience," without "Einstein."[10]

Yet while it's hard to argue with vague, positive-sounding ideas like "collaboration," what lies beneath such honorifics are very large claims about the role of big data and AI driving the future of neuroscience. Markram believes, apparently, that sheer data (and deep learning) will drive theory formation, yet the entire history of science and the short but explosive history of the big data fad both demonstrate the folly of such an approach.

We've already seen that successes in big data promoted in popular accounts (like Mayer-Schönberger and Cukier's *Big Data*) stem mainly from social domains like business, where no robust theories undergirding behavior are known to exist. In the absence of theory, Big Data AI has been a boon to many areas of interest where, without data-intensive methods, no real predictive progress could be made. Such examples may be encouraging for business leaders and may even illuminate interesting areas of popular culture, but they are generally inadequate and even disastrous for serious science.

We've seen how theory operated in supposed tech triumphs like the Higgs boson discovery. In such cases a robust theoretical framework makes possible a set of controlled and focused experiments that can help confirm predicted results of theory. Science has always allied itself with experiments in this sense, so the Higgs case is hardly controversial. But neither does it do much to buttress radical claims of the powers of Big Data AI along the lines that Markram would require in neuroscience.

Again, the lack of a robust theoretical framework in neuroscience makes an approach centered on data and machine learning even more misguided. The fewer theories researchers have to guide data-driven efforts, the fewer well-defined hypotheses can be tested, and the more such efforts will fall victim to known weaknesses in data-driven approaches.

Michael Jordan, an IEEE Fellow and the Pehong Chen Distinguished Professor of Computer Science at the University of California, Berkeley, has argued against those who see a linear connection between big data and scientific thinking (by which the more data we have, the better our scientific thinking becomes). As one of the world's most respected authorities on machine learning and big data, Jordan is an unlikely critic, but he predicts that "society is about to experience an epidemic of false positives coming out of big-data projects." As he puts it: "When you have large amounts of data, your appetite for hypotheses tends to get even larger. And if it's growing faster than the statistical strength of the data, then many of our inferences are likely to be false. They are likely to be white noise."[11]

For any particular data, Jordan argues, "I will find some combination of columns that will predict perfectly any outcome, just by chance alone. . . . I will find all kinds of spurious combinations of columns, because there are huge numbers of them. So it's like having billions of monkeys typing. One of them will write Shakespeare."[12]

Jordan here is pointing to the well-known problem in statistics known as overfitting (discussed below). Depressingly for Markram and other advocates of Data Brain projects like the Human Brain Project, overfitting is particularly problematic in the absence of causal or theoretical information about a domain—in the absence of general intelligence, that is. "Correlation is not causation" is a familiar caution, but bold claims made on behalf of Big Data AI projects in recent years have made it particularly relevant again. Apparently obvious truths about knowledge are seemingly now in need of restatement in the wake of major claims made about data and machine learning. As Marcus and Davis were forced to point out in the *New York Times,* "a big data analysis might reveal, for instance, that from 2006 to 2011 the United States murder rate was well correlated with the market share of Internet Explorer: Both went down sharply. But it's hard to imagine there is any causal relationship between the two."[13]

This is an obvious point, and it applies with force to hyped claims about deep learning. Buzzwords have changed, but the major fad pushing Big Data AI as a panacea threatens progress in fundamental areas like neuroscience, in spite of bold claims made by enthusiasts like Markram and others. The takeaway here is that the myth really does have practical consequences for our human futures—in actual science.

OVERFITTING

Statistician Nate Silver has also pointed out the inherent danger of overfitting theories (models) to data, where "overfitting" here means spuriously matching a set of data points to a description that contains no genuine explanatory power, because the description does not generalize to new, unseen data points in the underlying distribution in question. Generalization means abstracting away from irrelevant details of data and isolating the genuine relationships in a principle representation, or theory.

The simplest case of a "theory" or model of a set of data points is the linear interpolation of a scatter plot. Each data point on the coordinate system could be traced with a more complicated line that describes the existing points, but the description would be useless for new points plotted, as no actual knowledge of the point distribution is contained in such a description. A straight line rather shows the average or linear interpolation of the scattered data, and as such gives us a useful model predicting the behavior of the data. Overfitting, as Silver points out, gives false confidence on existing data, but quickly shatters this illusion when new data arrive—and don't fit the model or theory.[14]

Overfitting is a known problem in statistical inference or analysis, and it is frequently the cause of major failures in large scientific efforts as well. Here, again, the availability of theory can help researchers steer clear of over-fitted models and spurious correlations. As Silver notes, several high-profile attempts to predict earthquake occurrences (using historical data about earthquakes as well as detailed geographical information about stresses and other phenomena occurring under the Earth's surface along fault lines) failed miserably, though they perfectly fit existing data about earthquakes.

Like the meandering line explaining the existing points on a scatter plot, the models turned out to have no predictive or scientific value. There are numerous such fiascoes involving earthquake prediction by geologists, as Silver points out, culminating in the now-famous failure of Russian mathematical geophysicist Vladimir Keilis-Borok to predict an earthquake in the Mojave Desert in 2004 using an "elaborate and opaque" statistical model that identified patterns from smaller earthquakes in particular regions, generalizing to larger ones.

Keilis-Borok's student David Bowman, who is now Chair of the Department of Geological Sciences at Cal State Fullerton, admitted in a rare bit of scientific humility that the Keilis-Borok model was simply overfit.

Bowman went on to explain that earthquake prediction is limited because a theoretical understanding of what is happening under the Earth's surface along fault lines is lacking. Absent a genuine theory to use in guiding statistical or data-driven approaches, models are in constant danger of picking up "noise" rather than "signal," explains Silver.

NEUROSCIENCE: REFUSING TO LEARN FROM OTHERS' MISTAKES?

These lessons seem particularly germane in neuroscience today, as Big Data efforts seem destined to repeat mistakes made in other areas of science. Tellingly, the focus on Big Data AI is neither particularly new when looked at in this way, nor particularly encouraging. Theory in science, one might say, can never really be eliminated. One irony of the inevitability myth here is that theory is necessary not only for genuine science, but also for making good on dreams of general intelligence in AI. Modern confusions and mythology has the tail wagging the dog.

Yet, there are in fact existing theories about the spiking activity of neurons and the role of smaller function units like neuronal circuits in regions of the brain. There are even high-level theories of cognition or intelligence grounded in the workings of the human neocortex. What is lacking, as Markram and others point out, is a unifying framework or theory that fits these disparate pieces together.

The current Data Brain efforts are committed to the idea that progress is made possible by machine learning and AI in an essentially ground-up fashion. This is difficult to take seriously for at least two reasons. One, the high-level neocortex inspired theories of intelligence that animate much of the vision of Data Brain projects reproducing the human mind in silica are hopelessly general and unusable. The theories themselves are of very little use (ironically) to computer science or Artificial Intelligence engineering efforts, as they don't tell us enough about what the brain is actually doing when it generates intelligent behavior. Hence existing high-level theories already suggest that data-driven efforts are providing us with a pale and too-general set of theoretical postulates to begin.

Charitably, Markram no doubt believes that ground-up progress made by simulating larger and larger functional units of the brain will somehow improve or complete neocortical models of cognition. A better conclusion is that the focus on Big Data AI seems to be an excuse to put forth a number of vague and hand-waving theories, where the actual details and the ultimate success of neuroscience is handed over to quasi-mythological claims about the powers of large datasets and inductive computation. Where humans fail to illuminate a complicated domain with testable theory, machine learning and big data supposedly can step in and render traditional concerns about finding robust theories otiose. This seems to be the logic of Data Brain efforts today.

To see first-hand the overly general and relatively poor state of existing high-level theories, we turn next to a survey of neocortex-inspired theories of human intelligence. Far from demonstrating the continued place of theory in neuroscience, they show rather an increasing willingness to conform theories coming from neuroscience to the computing paradigm popular today.

This suggests that the weaknesses of such theories leave visionary neuroscientists like Markram undeterred in large part because of the prevalent belief that the march of Big Data AI toward general intelligence and beyond will fill in the details later. Rather than challenge the science that is occurring in neuroscience today, Data Brain advocates are increasingly willing to hand off mysteries and weaknesses to the supposed magic of data AI.

It's to these neocortical theories we turn next.

Chapter 17

...

NEOCORTICAL THEORIES OF HUMAN INTELLIGENCE

A popular theory of intelligence has been put forth by computer scientist, entrepreneur, and neuroscience advocate Jeff Hawkins. Famous for developing the Palm Pilot and as an all-around luminary in Silicon Valley, Hawkins dipped his toe into the neuroscience (and artificial intelligence) waters in 2004 with the publication of *On Intelligence,* a bold and original attempt to summarize the volumes of neuroscience data about thinking in the neocortex with a hierarchical model of intelligence.[1] He has since formed a company, Numenta, dedicated to unlocking the secrets of intelligence as computation.

The neocortex, Hawkins argues, takes input from our senses and "decodes" it in hierarchical layers, with each higher layer making predictions from the data provided by the lower ones, until the top of the hierarchy is reached and some overall predictive theory is synthesized from the output of all the lower layers.

His theory makes sense of some empirical data, such as differences in our responses based on different types of input we receive. For "easier" predictive problems, the propagation up the neocortex hierarchy terminates sooner (because the answer becomes available), and for tougher problems, the cortex keeps processing and passing the neural input up to higher, more powerful, more globally sensitive layers. The solution is then made available or passed back to lower layers until we have a coherent prediction based on the original input.

KURZWEIL'S HIERARCHICAL-PATTERN RECOGNITION THEORY

The hierarchical nature of the neocortex has also been noted by Ray Kurzweil, who writes in his 2012 *How to Create a Mind* that "the neocortex is responsible for dealing with our ability to deal with patterns and to do so in a hierarchical fashion. Animals without a neocortex (basically non-mammals) are largely incapable of understanding hierarchies."[2] Kurzweil credits his own hierarchical theory, which he calls the pattern recognition theory of mind, as working off accepted neuroscience findings on the structure and function of the human neocortex, and also off progenitors like Hawkins's own hierarchical account.

The hierarchical structure of the neocortex is indeed well-founded neuroscience. The columnar organization of the neocortex was first discovered by American neuroscientist Vernon Mountcastle in 1957. Mountcastle noted that the neocortex—a 2.5-millimeter-thick layer of nerve fibers stretching over the brain—was composed of columns of neurons, every one apparently identical to the others. There are about a half-million such columns in the human neocortex, each containing about sixty thousand neurons.

Kurzweil has hypothesized that each cortical column contains what he calls pattern recognizers, consisting of about one hundred neurons, totaling about 300 million pattern recognizers in the human neocortex. Like Hawkins, Kurzweil views these hypothetical recognizers as arranged into hierarchies that are centrally responsible for the unique capabilities of human thinking.

It's an interesting hypothesis. Both Hawkins and Kurzweil make the mistake, however, of believing that human intelligence is simple.

Earlier in these pages I have pointed out the facile nature of such theories of human intelligence, and I am not alone in noticing the hopeless generality of such attempts at fundamental theory, which seem more suited to propound particular computer system architectures of interest to the authors than to produce real understanding of the human brain. As Gary Marcus has pointed out, like Hawkins, Kurzweil seems to offer a hand-waving theory of AI based on vague insights about the brain. Says Marcus: "We already know that the brain is structured, but the real question is what all that structure does, in technical terms. How do the neural mechanisms in the brain map onto the brain's cognitive mechanisms?"[3]

Marcus goes on to point out that such theories are entirely too generic to advance the ball in neuroscience and related efforts in AI and the cognitive sciences: "Almost anything any creature does could at some level be seen as hierarchical-pattern recognition; that's why the idea has been around since the late nineteen-fifties. But simply asserting that the mind is a hierarchical-pattern recognizer by itself tells us too little: it doesn't say why human beings are the sort of creatures that use language (rodents presumably have a capacity for hierarchical-pattern recognition, too, but don't talk), and it doesn't explain why many humans struggle constantly with issues of self-control, nor why we are the sort of creatures who leave tips in restaurants in towns to which we will never return."[4]

Such generic theories are, ironically, also inspired by Big Data AI in a roundabout but very real way. Kurzweil is known for using hierarchical methods in machine learning for speech recognition applications; he worked on the original Siri application now owned by Apple and part of the iPhone. Hierarchical hidden Markov models are part of the data-analytic techniques that have merged with big data. And more recently, the now ever-present deep learning networks are arranged in hierarchies of layers. All such methods use large datasets as input to learn patterns in data, inducing a binary model that can then be decoded on unseen data.

Indeed, hierarchical learning methods today are almost as trendy as big data itself was a decade ago—witness deep learning. Theories in neuroscience, in other words, like entire projects such as the Human Brain Project and the BRAIN initiative, are becoming indistinguishable from methods that have found success in the computer sciences and in Big Data AI specifically.

MARKRAM'S LEGOS THEORY

Henry Markram is known for another overly general theory of learning, colloquially known as his "Legos theory of cognition," again basing the architecture on general research findings about the columnar and hierarchical nature of the human neocortex.[5] Markram and his coauthor Rodrigo Perin explain that, in this theory, "acquiring memories is very similar to building with Lego. Each assembly is equivalent to a Lego block holding some piece of elementary innate knowledge about how to process, perceive, and respond to the world."[6]

Again, an interesting hypothesis. Again, far too simple, and far too mechanistic.

We are now in a position to state explicitly what has become obvious. Not only have Data Brain efforts championed big data as a means to complete missing pieces of our understanding about the brain—as in, for instance, Markram's emerging principles of the connectome (synaptic connectivity)—but important theories themselves seem tied to computer science paradigms in such a way that perhaps the only meaningful direction Data Brain projects can now take is toward explicitly computational ideas and theories.

DEAD-END RESEARCH

We have seen that Big Data AI is not well-suited for theory emergence. On the contrary, without existing theories, Big Data AI falls victim to overfitting, saturation, and blindness from data-inductive methods generally. We can add here that data-centric computation has thus far produced quite poor and uninteresting theories that are suspiciously tied to popular technology approaches today.

This is not only Big Data AI but, of course, Big Science, as neuroscientists are beginning to realize. Paul Mathews, in his previously mentioned 2013 *Nature* interview, perhaps put it best: "I cannot think of major new conceptual advances that have come from such big science efforts in the past."[7] Markram and others committed to the Human Brain Project and Data Brain projects generally have pinned their hopes on Big Data AI to advance neuroscience, but what's really needed—just as Mathews suggests—is wide-ranging and disparate research agendas to encourage creative hypotheses and spur discovery. Big Data AI is not well-suited to these objectives.

Within a year of launch of the Human Brain Project, Markram and his vision drew intense criticism from a growing number of neuroscientists. In July 2014, more than five hundred scientists petitioned the European Commission to make major changes to the project, raising a number of concerns, many related to the project's faith in computation and big data at the expense of needed theory and creative research.

Ironically, the petition filed to the EU was in part a reaction to Markram's decision to shut down the cognitive architectures division of the project—the team specifically dedicated to exploring questions of cognition and intelligence, in line with Markram's stated broader vision. Neuroscientists also worried that the Human Brain Project did not set out to test any specific hypothesis or collection of hypotheses about the brain.[8]

The neuroscientists pointed out in the petition that more detailed simulations of the brain don't inevitably lead to better understanding. Hundreds of them, in other words, pushed back against the original Human Brain Project because it was not really neuroscience research at all, but rather a Big Data AI engineering project. Markram soon stepped down, but the project was retooled as software engineering—arguably less infected with AI mythology, but toothless for fundamental research by design.[9]

As Columbia University neuroscientist Eric Kandel put it, referring to the United States' BRAIN Initiative when it first launched, "We knew the endpoint [for the Human Genome Project] . . . . But here, we don't know what the goal is. What does it mean to understand the human mind? When will we be satisfied? This is much, much more ambitious."[10]

When the ten-year anniversary of Markram's now notorious TED talk proclaiming that our brains would be mapped into a supercomputer—the ultimate statement of mythology about AI—came in 2019, *Scientific American* (no enemy of future ideas about science) and *The Atlantic* both published searching accounts of what went wrong.[11] As one scientist put it, "We have brains in skulls. Now we have them in computers. What have we learned?"[12]

The questions were all about the dearth of theory. And no wonder. Data Brain enthusiasts like Markram seem to think that big data and the machine learning systems that analyze it will somehow provide answers to the questions we have about ourselves, about the human insight and intelligence that set those systems in motion.

This faith is not novel science but simply bad science, without a rich environment for future discovery.

Chapter 18

• • •

THE END OF SCIENCE?

Although scientists in growing numbers are discontented with Data Brain solutions to ongoing theoretical concerns, the ethos of Big Data AI is now firmly entrenched in science and culture generally. Ironically, as general intelligence is supposed to be emerging from AI and its applications to scientific research, it's noticeably downplayed in the roles of scientists. Billionaire tech entrepreneur and investor Peter Thiel remarked recently that innovations seem to be drying up, not accelerating.[1] Tech startups once dreamed of the next big idea to woo investors in the Valley, but now have exit strategies that almost universally aim for acquisitions by big tech companies like Google and Facebook, who have a lock on innovation anyway, since Big Data AI always works better for whoever owns the most data. The fix is in.

The question is whether, as Thiel puts it, there is now a "derangement of the culture," or whether the good ideas have already been snatched up.[2]

MEGABUCK SCIENCE

The polymathic MIT computer scientist and founder of cybernetics Norbert Wiener warned about what he called "megabuck" science in an unpublished manuscript, "Invention: The Care and Feeding of

Ideas," found among his papers after his death in 1964.[3] In the early 1950s, Turing had completed his fundamental (and it turned out, final) turn toward the future of invention as human-level AI; Wiener during the same period had begun serious contemplation about a future bereft of ideas necessary for AI, and other fields. Megabuck science emerged quickly in the aftermath of two world wars, with the Manhattan Project producing the atomic bomb, and major well-funded efforts underway on computer and communications theory and infrastructure. There were, for instance, efforts at Bell Labs and IBM as well as at large defense contractors like Raytheon. Modern science enjoyed an unprecedented history of significant and largely unpredictable invention—yet by mid-century, scientific innovation had become bureaucratized and controlled by large funding sources such as the US Department of Defense and major corporate interests. Wiener worried that, at the very moment of its triumph (and need), Western culture was turning toward downstream projects that ultimately threatened a flourishing culture of ideas.

His early 1950s manuscript (since published in 1993) now seems prophetic in its lament. "I consider that the leaders of the present trend from individualistic research to controlled industrial research are dominated, or at least seriously touched, by a distrust of the individual which often amounts to a distrust in the human."[4]

Wiener diagnosed megabuck science not only as suboptimal for a culture of invention, but as moving directly, and indeed happily, toward what he called an "antihuman" trend. This sentiment is echoed in our time by AI critics like Jaron Lanier, who worry about the tech-inspired erosion of personhood. Hive minds and swarm science would do little to quell Wiener's worries about the direction of science. As Wiener put it, "The general statistical effect of an anti-intellectual policy would be to encourage the existence of fewer intellectuals and fewer ideas."[5] Such anti-intellectual policies are so clearly evident in modern data-centric treatments of science that the threat is now impossible to ignore.

Wiener pointed out what we all know, or should know, which is that ideas emerge from cultures that value individual intellects: "New ideas are conceived in the intellects of individual scientists, and they are particularly likely to originate where there are many well-trained intellects, and above all where intellect is valued."[6]

It would indeed take a derangement in culture not to recognize the wisdom of Wiener's comments, which should be entirely uncontroversial. While lip service is given to brilliant innovation today just as in the 1950s, the reality is far different. The culture has become, as Wiener worried, sanguinely anti-intellectual and even antihuman.

The connection here to the myth is unavoidable, as mythology about the coming of superintelligent machines replacing humans makes concern over anti-intellectual and anti-human bias irrelevant. The very point of the myth is that anti-humanism is the future; it's baked into the march of existing technology.

It's difficult to imagine a cultural meme that is more directly corrosive to future flourishing and, paradoxically, more directly inimical to the very invention or discovery of a workable theory of general intelligence. Whether such a theory is forthcoming in future research and development is itself an unknown, but what can be recognized is the threat of an increasingly anemic culture of ideas that will militate against any such discovery. The overall effect of the myth in this context is simply to push AI, and indeed scientific research itself, into a techno-centric mode, where genuine invention will be systematically discouraged and go unrecognized—if, as is always rare in all ages, and even more so today, it actually occurs.

BETTING ON IDEAS

Wiener pointed out that the economics of corporate profit make investment in a genuine culture of ideas difficult, since early bets on ideas are all in essence bad, as their full value becomes apparent only downstream.

To put it simply, new ideas can't be predicted, and so represent an economic and intellectual commitment to a flourishing culture without guaranteed short-term gain. We should expect, in other words, that the consolidation of the web into big tech will also tend to skew work on AI toward narrow applications on the profit curve, while inventions (still unknown) get short shrift.

As proof of this claim, consider how little investment is given to exploring paths to artificial general intelligence, as opposed to applications of, for instance, deep learning to gameplay. The latter is clearly a dead-end to artificial general intelligence, as even deep learning researchers are now beginning to admit—wary as they no doubt are of another notorious AI winter on the heels of a new bubble. The culture is squeezing profits out of low-hanging fruit, while continuing to spin AI mythology, a strategy guaranteed to lead to disillusionment without an inflow of radical conceptual innovation.

Wiener wryly observed that Swift's farcical *Laputa* world, where a machine evolves science "automatically," had a certain intellectual footing in 1950s megabuck science, with the inevitable result of further pushing away a culture of invention. He was particularly worried about early versions of what is now part and parcel of AI mythology, that the human mind is getting replaced by computer programs: "the present desire for the mechanical replacement of the human mind has its sharp limits. Where the task done by an individual is narrowly and sharply understood, it is not too difficult to find a fairy adequate replacement either by a purely mechanical device or by an organization in which human minds are put together as if they were cogs in such a device."[7]

Wiener's remark is, of course, a perfect restatement of AI mythology and its deleterious effect on humanity, with hive minds on the web and swarm science in scientific research. We might be forgiven for not "waiting around" for invention and discovery while we have IBM Blue Gene supercomputers to play with, but what's unforgivable is the deliberate attempt to reduce personhood, as Lanier puts it—disparaging and taking away the importance and value of the human mind itself. Such a strategy is fantastically self-defeating and stupid.

Wiener then connected his critique to popular mechanistic fancies, often lampooned by skeptics ever wary of machine dreams. (We saw Jonathan Swift's farce of mechanical science appear earlier, in Peirce's discussion of early developments in automated reasoning.) Wiener continued: "However, the use of the human mind for evolving really new thoughts is a new phenomenon each time. To expect to obtain new ideas of real significance by the multiplication of low-grade human activity and by the fortuitous rearrangement of existing ideas without the leadership of a first-rate mind in the selection of these ideas is another form of the fallacy of the monkeys and the typewriter, which already appears with a slightly simpler statement in Swift's *Voyage to Laputa*."[8]

Henry Markram's fantasy of turning a billion euros into AI mythology by building a brain using neural networks and supercomputers (and existing neuroscientific theories) is captured perfectly by Wiener here. If only these ideas had been exposed and avoided. In fact, the modern turn in AI seems to have pulled such fancies even more centrally into culture, with predictably narrow but flashy application successes touted as the future, which (alas) will be dominated by superintelligent machines. The vision of artificial general intelligence here is pure mythology and window dressing. No one is likely to understand even the core problems clearly, let alone happen upon the ideas necessary for true progress. This comparison might invite a smirk, but it's nonetheless apt: it's a brave new world. Wiener, much to his credit, saw it coming.

NARROWER AND NARROWER

Turning to AI in the inference framework, we are witnessing in effect the evolution of a sub-species in inductive AI, which can perform well in narrow, data-centric environments but necessarily lacks the ability to learn common sense and acquire genuine understanding. That we are pinning the future of the human mind—not so constrained—on the further development of AI in this vein is simply stupefying.

Not only is this approach entirely bereft of the general intelligence necessary to make any real intellectual advance in modern culture, but because induction is provably distinct from abduction, we already know that there is no bridge from the one to the other. All of Ray Kurzweil's proclamations of inevitable progress cannot undo this truth once it becomes known. We should be honest here, as recognition of the truth would itself form part of the blueprint for moving forward.

To sum up: there is no way for current AI to "evolve" general intelligence in the first place, absent a fundamental discovery. Simply saying "we're getting there" is scientifically and conceptually bankrupt, and further fans the flames of antihuman and anti-intellectual forces interested in (seemingly) controlling and predicting outcomes for, among other reasons, maximizing short-term profit by skewing discussion toward inevitability. Smart individuals change the course of things; one way to make the future more predictable is simply to disparage and eliminate any value placed on individual intelligence.

MOVE ALONG—THERE'S NOTHING TO SEE HERE

The suggestion that we've wandered into a cultural dead end might seem fantastic and fictional if in fact many of the purveyors of AI mythology weren't happily on record pooh-poohing Wiener's "care and feeding of ideas" concerns, while talking up the inevitability of AI. While AI scientists and part-time mythologists like Stuart Russell still admonish us not to discount human ingenuity in the pursuit of a future theory of artificial general intelligence, very few leaders in the current culture are actually pursuing an agenda where human ingenuity can thrive.

Given the expressed aims (or fears) of creating in effect a new super-being, this is astounding. Surely we could use an Einstein or two these days. (One wonders how Turing would fare today.)

Again, nowhere is this more evident than in the dogma of AI mythology itself. On any calculation about the future of artificial general intelligence, the onus is squarely on AI mythologists portending the coming of human-level AI to explain what we're doing to move things along.

Perhaps we could start with a frank acknowledgement that deep learning is a dead end, as is data-centric AI in general, no matter how many advertising dollars it might help bring in to big tech's coffers. We might also give further voice to a reality that increasing numbers of AI scientists themselves are now recognizing, if reluctantly: that, as with prior periods of great AI excitement, no one has the slightest clue how to build an artificial general intelligence.

The dream remains mythological precisely because, in actual science, it has never been even remotely understood. Where else but in AI science itself should we get rid of the myth?

JOHN HORGAN AND THE DISQUIETING SUGGESTION OF THE END

The specter of a purely technocratic society where science, which once supplied us with radical revolutionary discoveries and inventions, now plays the role of lab-coated technician tweaking knobs on the "giant brains" of supercomputers, was suggested early on by *Scientific American* writer John Horgan. In his hugely popular *The End of Science,* Horgan in the mid-1990s wondered whether the seeming petering out of basic research in science was inevitable, for the simple fact that major discoveries are behind us.[9]

This is one half of Thiel's question today: is the culture deranged, set on a course of choking out new ideas, as Wiener worried, or are we actually out of basic ideas, because we've already found them all? This latter possibility would represent "The End" in a basic sense—so we might pray that culture has merely embraced an all-encompassing technological answer to basic questions that is only asphyxiating human intelligence as a by-product. There is at least a hypothetical way to fix a deranged culture of science; escape from the *Tron*-world of the end of ideas represents a further nightmare.

Thiel's question is central to the future not just of AI but of humanity, and unfortunately we have evidence for both hypotheses. On the one hand, cheerful promotion of the myth, and its cousin in swarm science—like cheerleading for hive minds before it—seems to suggest that modern society has indeed wandered whistling into a kind of derangement of core values, precisely as Wiener portended.

On the other hand, the question of whether we have no choice, as Horgan argues, presents a disquieting possibility that now, more than three hundred years after the Scientific Revolution, all the low-hanging fruit of physical and computational theory have been picked. In this view, we've already discovered more or less what could be discovered about physics, with first Newton's laws and then Einstein's relativity and the development of twentieth-century quantum mechanics. Remaining physics progress will be largely about filling in gaps and details in existing theory, and no doubt testing the predictions of such theories with larger and more expensive technologies like supercolliders. Welcome to Machineland.

Either negative possibility would support Markram's suggestion that Einstein is now unwanted, and doesn't have anything left to do today (except contribute to data science). The inevitability of a coming superintelligence is here turned upside-down, because humans, so brilliant in discovering the fundamental building blocks of the universe, now must retire and watch as the culture turns from discoverers to technicians. Tending supercomputers is the modern version of Voltaire's tending the garden. The serious work is over. Human beings shouldn't have been so smart.

Horgan also suggests that some dreams, like the dream of a complete scientific account of human consciousness, might be too difficult anyway, and impossibly remote. In this case, we have the unhappy result of witnessing the inexorable creep of computation—computing existing theories—into science and everywhere else, while Promethean dreams of a completed neuroscience are quietly put to bed, or fictionalized in *Ex Machina* futures.

It certainly is possible that major scientific advance is behind us, in which case we should expect shallow technical treatments of core issues, using what existing theories we have, while AI mythology becomes a new focal point for future meaning, however nihilistic and untrue. As Lanier suggested, too, we can make such a future become true by simply chiseling away at human intelligence and uniqueness, until we have stooped low enough to adjust to a computation-dominated future.

Horgan was not thrilled with his own disquieting suggestion, but it does seem that, since the 1990s, applied computation has lent ever-increasing credibility to it—if not in reality, then at least in observed practice.

In either case, we should take seriously that we are now on the wrong path, in large part because we are actively attempting to cover up a key deficiency—a lack of flourishing human culture—with rhetoric about the inevitable rise of machines. Eugene Goostman could not have come up with a better path to non-achievement.

OUR CHOICE

If Horgan's "The End" reading of our future is true, the drift into technical detail is inevitable. Yet a derangement of culture spread in large part by the myth (and the rise of ubiquitous computation) keeps alive the possibility that freeing ourselves of modern technology myths might spur progress by causing reinvestment in human insight, innovation, and ideas.

Clearly, I favor the latter interpretation. And I'm optimistic—largely because, as we've seen, on purely scientific grounds we have every reason to reject a linear and inevitable march to artificial general intelligence (and beyond).

Untying this Gordian knot starts with ridding ourselves of the myth in its current guise, which has infected culture so pervasively that long discussions about the need for new theory in neuroscience are now required to refocus efforts—a point that should be clear and in need of no argument.

TRUST AS RECOGNITION OF LIMITS TO INDUCTIVE SYSTEMS

Ironically, the limits of modern AI are implicit in current discussions about automation and trust. It has become trendy for AI thinkers to worry about so-called "beneficial AI," trusted systems, and other ethical issues like problematic bias. In other words, systems that don't understand but still perform have become a concern.

This cuts the myth at an awkward angle: it is because the systems are idiots, but still find their way into business, consumer, and government application, that human-value questions are now infecting what were once purely scientific issues.

Self-driving cars are an obvious case in point. It's all well and good to talk up advances in visual object recognition until, somewhere out on the long tail of unanticipated consequences and therefore not included in the training data, your vehicle happily rams a passenger bus as it takes care to miss a pylon. (This happened.) Look, too, at the problems with bias and image recognition: Google Photos slapped a GORILLA label on a photo of two African-Americans. After that neutron bomb of a PR disaster, Google fixed the issue—by throwing images of gorillas out of the training set used by the deep learning system.

Thus limits to inductive AI lacking genuine understanding are increasingly pushed into AI discussion because we are rushing machines into service, in important areas of human life, which have no understanding. This, too, is a consequence of AI mythology, which shows a continuing penchant for not waiting around for legitimate ideas or discoveries, only too eager to keep increasing the dominion of AI technologies in every possible area of life.

Ironically, this worrisome trend could help spur more understanding of AI's fundamental—or at the very least, current and unavoidable—limits. Actual human lives and important human values are now at issue.

In the name of the myth, in other words, much ink is spilled today describing what amounts to the stupidity of machines. No one seems to notice that the result is a necessary and foreseeable consequence of inductive systems masquerading as a path to intelligence.

Russell points to the "alignment" problem, an issue in AI of suddenly central importance, concerned with aligning current and future AI systems with our own interests and purposes. But the problem arises not, as Russell suggests, because AI systems are getting so smart so quickly, but rather because we've rushed them into positions of authority in so many areas of human society, and their inherent limitations—which they've always had—now matter.

I'm hopeful that the current turn away from the Singularity toward practical concerns about ceding real authority to AI—to, let's face it, mindless machines—will eventually result in a renewed appreciation for human intelligence and value.

Considering the alignment problem might give rise to considerations of augmentation—how we can best use increasingly powerful idiots savants to further our own objectives, including in the pursuit of scientific progress.

IN CONCLUSION

The inference framework I've presented in this book clarifies the project of expanding current artificial intelligence into artificial general intelligence: it must bridge to a distinct type of inference, currently not programmable. It also provides a guide to exploring boundaries between minds and machines that can facilitate more optimal and safer human-machine interactions, which are of course here to stay. It remains true that technology often acts like a prosthetic to human abilities, as with the telescope and microscope. AI has this role to play, at least, but a mythology about a coming superintelligence should be placed in the category of scientific unknowns. If we wish to pursue a scientific mystery directly, we must at any rate invest in a culture that encourages intellectual ideas—we will need them, if any path to artificial general intelligence is possible at all.

Just as *Frankenstein* was really an exploration of spiritual isolation (a problem felt deeply by Mary Shelley and her husband, Percy Shelley), the deepest questions embodied in the AI myth are not technical or even scientific—they involve our own ongoing attempts to find meaning and to forge future paths for ourselves in an ever-changing world. There is nothing to be gained by indulging in the myth here; it can offer no solutions to our human condition except in the manifestly negative sense of discounting human potential and limiting future human possibility.

The problem of inference, like the problem of consciousness, is entrenched at the center of ongoing grand mysteries, and is really presupposed in our understanding of everything else. We should not be surprised that the undiscovered mind resists technological answers. It's possible that, as Horgan worried, we're out of ideas. If so, the myth represents our final, unrecoverable turn away from human possibility—a darkly comforting fairy tale, a pretense that out of our ashes something else, something great and alive, must surely and inevitably arise.

If we're *not* out of ideas, then we must do the hard and deliberate work of reinvesting in a culture of invention and human flourishing. For we will need our own general intelligence to find paths to the future, and a future better than the past.

NOTES

Introduction

1. I don't mean to suggest that researchers have not wrestled with abduction in AI—they have. In the 1980s and 1990s researchers worked on logical approaches to abduction, called abductive logic programming. But these systems were abduction "in name only," because they relied on deduction, not true abduction. The systems weren't successful, and were quickly abandoned as work in AI progressed in the web era. More recently, circa 2010 up to present day, various probabilistic (in particular, Bayesian) approaches have been adopted as possible routes to bona fide abductive inference. Those systems, however, are not full treatments of abduction either. Instead of disguised deductive approaches like their predecessors, they are disguised inductive or probabilistic approaches. Abduction in name only is not what I mean by abduction, and the systems using the name but not solving the problem won't help us make progress in AI. I will explain all this in pages to come.

Chapter 1: The Intelligence Error

1. A. M. Turing, "Computing Machinery and Intelligence," *Mind* 59, no. 236 (October 1950), 433–460.

2. A. M. Turing, "On Computable Numbers, with an Application to the Entscheidungsproblem," *Proceedings of the London Mathematical Society*, vols. 2–42, issue 1 (January 1937), 230–265.

3. A. M. Turing, *Systems of Logic Based on Ordinals* (PhD diss., Princeton University, 1938), 57.

4. Gödel also showed that adding rules would patch up incompleteness in some systems, but that the new system, with the additional rules, would have yet other blind spots, on and on. This was precisely Turing's focus in his later work on formal systems and completeness.

5. For the original incompleteness results see Kurt Gödel, "Über formal unentscheidbare Sätze der Principia Mathematica und verwandter Systeme I," *Monatshefte für Mathematik Physik* 38 (1931): 173–198. An English translation is in Kurt Gödel, *Collected Works*, vol. 1: 1929–1936, eds. Kurt Gèodel, Kurt Gödel, and Solomon Feferman (Oxford: Oxford University Press, 1986).

6. I am using the terms *formal, mathematical,* and *computational* interchangeably here. The terminology is not imprecise, although technically all mathematical or computational systems are known as formal systems. I trust this is not confusing, but in any case, the terms mathematical and computational both refer to formal systems, which have a well-defined vocabulary of symbols, and rules for manipulating the symbols. This covers arithmetic as well as computer languages, and is completely general to suit the purposes of the discussion.

Chapter 2: Turing at Bletchley

1. Turing, Good, and Shannon's early work on computer chess exploited a technique known as minimax, which scored moves based on minimizing loss for a player while maximizing potential gain. The technique figured prominently in later versions of computer chess, and is still a baseline for designing the much more powerful computer chess systems used today.

2. Full programming languages as we know them today, such as C++ and Java, all make use of these basic operations that emerged out of early computing, although ideas such as object oriented programming and other means of structuring software emerged later in computer science. Still, the basic control structures in all computer code appeared early on with the first full electronic machines. Insight into how to structure and control machines with programs is no doubt responsible for the immediate success of such systems applied to early problems like chess.

3. Under the direction of Britain's "General Code and Cipher School" or "GC and CS."

4. Hugh Alexander, for instance, was a national chess champion who helped in the Bletchley effort.

5. The Germans added rotors that generated long sequences of ciphered free text, along with a more complicated initial (and also ciphered) instructions for deciphering communications.

6. See Andrew Hodges's excellent biography of Turing for in-depth discussion of the role of Bletchley Park and Turing in World War II. Andrew Hodges, *Alan Turing: The Enigma* (New York: Vintage, 1992).

7. Joseph Brent, *Charles Sanders Peirce: A Life* (Bloomington, IN: Indiana University Press, 1993), 72.

8. Hodges, *Enigma*, 477.

9. François Chollet, "The Implausibility of Intelligence Explosion," *Medium*, November 27, 2017.

10. For a mathematical treatment of the no free lunch theorem, see David Wolpert and William G. Macready, "No Free Lunch Theorems for Optimization," *IEEE Transactions on Evolutionary Computation* 1, no. 1 (1997): 67–82.

11. The term artificial intelligence was actually coined in 1955 by Stanford computer scientist John McCarthy, one of the pioneers of AI and a member of the Dartmouth Conference which, a year later in 1956, officially launched the field.

Chapter 3: The Superintelligence Error

1. Irving John Good, "Speculations Concerning the First Ultraintelligent Machine," *Advances in Computers* 6 (1965) 6: 31–88.

2. Nick Bostrom, *Superintelligence: Paths, Dangers, Strategies*, repr. ed. (Oxford: Oxford University Press, 2017), 259.

3. John Von Neumann, *Theory of Self-Reproducing Automata*, ed. Arthur W. Banks (Urbana: University of Illinois Press, 1966), fifth lecture, 78.

4. Daniel Kahneman, *Thinking, Fast and Slow* (New York: Farrar, Straus and Giroux, 2013).

5. Stuart Russell, *Human Compatible: Artificial Intelligence and the Problem of Control* (New York: Viking, 2019), 37.

6. Kevin Kelly, *What Technology Wants* (New York: Penguin, 2010).

7. Strangely, or maybe refreshingly, Kelly has since distanced himself from the AI myth. Writing in *Wired* in 2017, he argues that "intelligenization" isn't leading to superintelligence after all. He points out that intelligence is varied and polymorphous, and that seemingly unintelligent animals

like squirrels remember the locations of potentially thousands of buried nuts for later consumption, a feat which humans would no doubt fail to replicate. The title of his piece, "The AI Cargo Cult: The Myth of Superhuman AI," is telling. Our inability to fix a workable definition of intelligence (let alone superintelligence) might suggest that the endpoint futurists see as inevitable in AI is actually confused, and yet another simplification that provides ample room for mythology and speculation.

8. Russell, *Human Compatible,* 7–8.

Chapter 4: The Singularity, Then and Now

1. Murray Shanahan, *The Technological Singularity* (Cambridge, MA: MIT Press, 2015), 233.

2. As we've seen, the science of Turing's universal machines was well established by the late 1930s. Computers as electronic devices appeared later, on the heels of developments of communications technologies such as relay switches from telephone systems and other technologies.

3. Technically, Vinge introduced the term singularity three years earlier, in a January 1983 article in *Omni* magazine titled "First Word." It is common, however, to trace the word and Vinge's use of it back to his sci-fi book *Marooned in Realtime,* where the concept was fully developed in the plot of the story.

4. Vernor Vinge, "The Coming Technological Singularity: How to Survive in the Post-Human Era," in *Vision-21: Interdisciplinary Science and Engineering in the Era of Cyberspace,* ed. G. A. Landis, NASA Publication CP-10129, 1993, 11–22.

5. Ray Kurzweil, *The Singularity is Near: When Humans Transcend Biology* (New York: Penguin Group, 2005).

6. Ray Kurzweil, "The Singularity: A Talk with Ray Kurzweil," interview with The Edge, introduction by John Brockman, March 24, 2001, https://www.edge.org/conversation/ray_kurzweil-the-singularity.

7. Hubert L Dreyfus, *What Computers Still Can't Do: A Critique of Artificial Reason* (Cambridge, MA: MIT Press, 1992), ix.

Chapter 5: Natural Language Understanding

1. John McCarthy, M. Minsky, N. Rochester, and C. E. Shannon, "A Proposal for the Dartmouth Summer Research Project on Artificial Intelligence," August 1955.

2. Gary Marcus and Ernest Davis, *Rebooting AI: Building Artificial Intelligence We Can Trust* (New York: Pantheon Books, 2019), 1.

3. Massimo Negrotti, ed., *Understanding the Artificial: On the Future Shape of Artificial Intelligence* (Berlin Heidelberg: Springer-Verlag, 1991), 37.

4. See John R. Pierce et al., *Language and Machines: Computers in Translation and Linguistics,* report of Automatic Language Processing Advisory Committee, National Academy of Sciences, National Research Council, Publication 1416, 1966.

5. Sergei Nirenburg, H. L. Somers, and Yorick Wilks, eds., *Readings in Machine Translation* (Cambridge, MA: MIT Press, 2003), 75.

6. For a readable discussion of early problems with machine translation, see John Haugeland, *Artificial Intelligence, The Very Idea* (Cambridge, MA: MIT Press, 1989). Yehoshua Bar-Hillel's comment appears on page 176.

7. See, for instance, Hubert Dreyfus's account of early automatic translation failures in Hubert L Dreyfus, *What Computers Still Can't Do: A Critique of Artificial Reason* (Cambridge, MA: MIT Press, 1992), ix.

8. For more on DENDRAL, see Robert K. Lindsay, Bruce G. Buchanan, E. A. Feigenbaum, and Joshua Lederberg, "DENDRAL: A Case Study of the First Expert System for Scientific Hypothesis Formation," *Artificial Intelligence* 61, no. 2 (1993): 209–261. For more on MYCIN, see B. G. Buchanan and E. H. Shortliffe, *Rule Based Expert Systems: The MYCIN Experiments of the Stanford Heuristic Programming Project* (Reading, MA: Addison-Wesley, 1984).

9. For a good discussion of the ELIZA program in work on Natural Language Processing in AI, see James Allen, *Natural Language Processing* (San Francisco: Benjamin / Cummings Publishing Company, 1995). The dialogue appears on page 7.

10. I explain the problems with the Goostman performance in much greater detail in Part Two.

Chapter 6: AI as Technological Kitsch

1. The term *technoscience* here is an anachronism, but retrospectively captures exactly the ideas taking shape in the nineteenth century in the wake of the Scientific Revolution. In fact, *technoscience* was coined in the 1970s by the Belgian philosopher Gilbert Hottois.

2. Friedrich Wilhelm Nietzsche, *The Gay Science; with a Prelude in Rhymes and an Appendix of Songs* (New York: Vintage Books, 1974).

3. Fyodor Dostoevsky, *Notes From Underground* (New York: Vintage Classics, 1994), 33.

4. Such views were also held by French thinkers who influenced Comte, including Marie Jean Antoine Nicolas Caritat, the Marquis de Condorcet (typically referred to simply as Condorcet). Prior to the revolution in French philosophy brought on by the Scientific Revolution, English philosophers such as Francis Bacon also espoused similar views, regarding science and progress as philosophically essential.

5. Hannah Arendt, *The Human Condition* (Chicago: Chicago University Press, 1958).

Chapter 7: Simplifications and Mysteries

1. B. F. Skinner, *Walden Two* [1948] (Indianapolis: Hackett, 2005).

2. Stuart Russell, *Human Compatible: Artificial Intelligence and the Problem of Control* (New York: Viking, 2019), 8.

3. Dan Gardner, *Future Babble: Why Expert Predictions Are Next to Worthless, and You Can Do Better* (New York: Dutton, 2011).

4. Martin Ford, *Architects of Intelligence: The Truth about AI from the People Building It* (Birmingham, UK: Packt Publishing, 2018), 20.

5. Ray Kurzweil, *The Singularity is Near: When Humans Transcend Biology* (New York: Penguin Group, 2005), 25.

6. Alasdair MacIntyre, *After Virtue* (Notre Dame, IN: University of Notre Dame Press, 2007), 111.

7. See Michael Polanyi, *Personal Knowledge: Towards a Post-Critical Philosophy* [1958] corrected edition (Abingdon-on-Thames: UK: Routledge & Kegan Paul, 1962), chapter 5.

8. Hubert Dreyfus, *What Computers Still Can't Do: A Critique of Artificial Reason* (Cambridge, MA: MIT Press, 1992).

9. Gary Marcus and Ernest Davis, *Rebooting AI: Building Artificial Intelligence We Can Trust* (New York: Pantheon Books, 2019).

10. Hector Levesque, *Common Sense, the Turing Test, and the Quest for Real AI* (Cambridge, MA: MIT Press, 2017).

11. Erik J. Larson, "Questioning the Hype About Artificial Intelligence," *The Atlantic,* May 14, 2015.

12. Stuart Russell, *Human Compatible: Artificial Intelligence and the Problem of Control* (New York: Viking, 2019), 9.

13. Russell, *Human Compatible,* 41.

14. Russell, *Human Compatible*, 16–17.

15. Ford, *Architects*, 232, 234.

16. The scene where Ava escapes Nathan's compound and looks up into the sunlight to see color addresses a philosophical problem known in AI and philosophy of mind circles as "Mary the color scientist." The conundrum is whether a fictional color scientist named Mary, who (by hypothesis) knows all the scientific facts about color (wavelengths of light impinging on neurons in the brain, and so forth) but lives in a black-and-white room, would actually learn something new when seeing actual color for the first time. In other words: Is our experience of seeing color additional to computations about it? Garland suggests that Ava did indeed learn something new. Presumably this also proves that she has a conscious mind.

17. Eliezer Yudkowsky, "Artificial Intelligence as a Positive and Negative Factor in Global Risk," in *Global Catastrophic Risks*, eds. Nick Bostrom and Milan M. Ćirković (New York: Oxford University Press), 308–345.

18. Jaron Lanier, *You Are Not a Gadget: A Manifesto* (New York: Alfred A. Knopf, 2010), 4.

Chapter 8: Don't Calculate, Analyze

1. Edgar Allan Poe, *The Best of Poe: The Tell-Tale Heart, The Raven, The Cask of Amontillado, and 30 Others* (Clayton, DE: Prestwick House, 2006).

2. Poe, *The Best of Poe*, 33–34.

3. Poe, *The Best of Poe*, 27.

Chapter 9: The Puzzle of Peirce (and Peirce's Puzzle)

1. Joseph Brent, *Charles Sanders Peirce: A Life* (Bloomington, IN: Indiana University Press, 1993), 1–7.

2. Quoted in Brent, *Peirce*, 2–3.

3. Henry James, ed., *The Letters of William James*, vol. 1 (Boston: Atlantic Monthly Press, 1920), 35.

4. Quoted in Brent, *Peirce*, 16.

5. Joseph Brent likens Peirce to a dandy: "The Dandy lives and sleeps in front of a mirror, is rich, and is consumed by his work, which he yet does in a disinterested manner. He is solitary and unhappy" (23–24). The novelist Henry James (brother of William) once quipped that Peirce was interesting and wore "beautiful clothes," after meeting a forlorn Peirce haunting Paris,

newly separated from his wife (Brent, *Peirce*, 25). Indeed, Peirce's life drew a litany of complaints by loved ones, colleagues, and superiors, who accused him of atheism (not true), alcoholism (perhaps true), drug addiction (true, but for a reason—he had a lifelong affliction of a painful condition known as facial neuralgia), marital infidelity (no doubt true), and recklessness in his professional affairs. The facts have never completely come out on his dismissal as lecturer at Hopkins. His inability to care for expensive gravity measuring equipment while employed by the US Coast Survey is well-documented, as is his tardiness on projects.

6. Brent, *Peirce*, 9.

7. Genuine scientific discoveries, almost by definition, tend to be improbable inferences (or constellations of inferences) when judged by prior accepted "evidence." After the discovery is accepted (if it is), what counts as evidence itself is reevaluated, along with judgments of likelihood.

Chapter 10: Problems with Deduction and Induction

1. Technically, universal bug-checking systems are impossible, as we know from Gödel's incompleteness theorems. But given this proviso, deductive systems are useful for checking whether software conforms to a set of specifications, among other things.

2. Wesley Salmon, *Causality and Explanation* (Oxford: Oxford University Press, 1998).

3. Here I am adopting the usual word in AI, "agents," which means anything that acts independently. An intelligent (or "cognitive") agent might be a person, but it could also be an AI system or an alien.

4. David Hume, *Hume's Treatise of Human Nature* [1739–1740], ed. L. A. Selby Bigge (Oxford: Clarendon Press, 1888), 89.

5. Russell's original formulation referred to a chicken, not a turkey. Karl Popper reformulated the example in its current guise. And the quote is taken from a secondary source: Alan Chalmers, *What Is This Thing Called Science?* 2nd ed. (St. Lucia, AU: University of Queensland Press, 1982), 41–42.

6. Stuart Russell, *Human Compatible: Artificial Intelligence and the Problem of Control* (New York: Viking, 2019), 48.

7. Gary Marcus and Ernest Davis, *Rebooting AI: Building Artificial Intelligence We Can Trust* (New York: Pantheon Books, 2019).

8. Marcus and Davis, *Rebooting AI*, 62.

9. Martin Ford, *Architects of Intelligence: The Truth about AI from the People Building It* (Birmingham, UK: Packt Publishing, 2018).

10. Nassim Nicholas Taleb, "The Fourth Quadrant: A Map of the Limits of Statistics," Edge.org, September 14, 2008, https://www.edge.org/conversation/nassim_nicholas_taleb-the-fourth-quadrant-a-map-of-the-limits-of-statistics.

11. Judea Pearl and Dana Mackenzie, *The Book of Why: The New Science of Cause and Effect* (New York: Basic Books, 2018).

12. I should say something more about the terms observation, data, and fact. Imagine a guy, Fred, who is tasked by his employer to collect data on everyone wearing a North Face jacket on a particular street corner in downtown Chicago on December 1, from 7:00 AM to 7:00 PM. Fred will observe people passing by, noting the jackets they are wearing. For each North Face jacket he sees, he will record the observation on a pad of paper. After 7:00 PM, he brings the pad of paper to his supervisor, who has the task of taking the recorded observations (the number 147), and inputting them into a spreadsheet. Fred's observations are now data—they have been recorded in a computer-readable format and represent the fact of the number of North Face jacket wearing pedestrians at the time and place. Hence, data are recorded observations, considered as facts. The whole issue of what a fact is actually quite interesting, but it will have to await some other project.

13. Apparently Pearl also has an interest in detective work, as he, too, points out that Sherlock Holmes didn't use deduction as our fictional hero sometimes claimed. He used induction, according to Pearl, in the sense of examining facts to arrive at their explanation. Certainly Pearl can't mean this literally, as the claim amounts to contradicting his own ladder schema, since associating observations without understanding would hardly crack the puzzles confronting Holmes. The conflation exposes an ongoing misapprehension about induction generally: that theories come from careful observation. This is true only if we allow that prior theory informs observation and that, literally speaking, observation alone can't supply missing theory. As we'll see with language understanding, induction enters as a part into a kind of holistic hermeneutical circle, where prior ideas are brought to interpretation, and ongoing interpretation keeps adjusting prior ideas and suggesting new ones. It's impossible to make sense of language relying on induction, correctly understood (that is, not packing into it other forms of inference).

14. Marcus and Davis, *Rebooting AI*.

15. Russell, *Human Compatible.*
16. Pearl, *The Book of Why*, 36.

Chapter 11: Machine Learning and Big Data

1. Stuart Russell, *Human Compatible: Artificial Intelligence and the Problem of* Control (New York: Viking, 2019).
2. Tom Mitchell, *Machine Learning* (New York: McGraw-Hill Education, 1997), 2.
3. We trust the filters in large part because they are deliberately permissive: spam is more likely to get into inboxes then legitimate messages are to get thrown out. The consequences of using systems with no guarantees and no real understanding are an increasing concern. Spam is a small concern, in comparison to self-driving cars.
4. Melanie Mitchell, *Artificial Intelligence: A Guide for Thinking Humans* (New York: Farrar, Straus, and Giroux, 2019).
5. Doug Laney, "3-D Data Management: Controlling Data Volume, Velocity and Variety," Gartner Group Research Note, February 2001. The original is no longer available from Gartner, but Laney has reposted it at https://community.aiim.org/blogs/doug-laney/2012/08/25/deja-vvvu-gartners-original-volume-velocity-variety-definition-of-big-data. IBM later coopted the catchy framework and added a fourth V, for veracity. But Laney then pointed out in a blog post (no longer accessible online) that veracity is actually inversely related to size, and therefore its addition messes up the Gartner definition. He explained that measures like veracity are not a measure of "bigness," and thus not a defining feature of big data more than once. Doug Laney, "Batman on Big Data," Garter Blog Network, November 13, 2013.
6. Gil Press, "12 Big Data Definitions: What's Yours?" *Forbes*, September 3, 2014.
7. See https://obamawhitehouse.archives.gov/blog/2016/05/23/administration-issues-strategic-plan-big-data-research-and-development.
8. Jonathan Stuart Ward and Adam Barker, "Undefined by Data: A Survey of Big Data Definitions," School of Computer Science, the University of Saint Andrews, UK. Published in arXiv, 2013.
9. Viktor Mayer-Schönberger and Kenneth Cukier, *Big Data: A Revolution That Will Transform How We Live, Work, and Think* (New York: Eamon Dolan / Mariner Books, 2014), 6.

10. Chris Anderson, "The End of Theory: The Data Deluge Makes the Scientific Method Obsolete," *Wired*, June 23, 2008.

11. Gil Press, "Big Data is Dead. Long Live Big Data AI," *Forbes*, July 1, 2019.

12. These can all be recorded as answers to yes / no questions in a data structure (called a vector) containing "1" for YES and "0" for NO. Thus {1,1,1,0,0,1} is a feature vector to be supplied as input to a learning algorithm (the "learner"). There are many ways to structure input for different learners—this is a simple example. Details don't matter as much as the general idea.

13. James Somers, "The Man Who Would Teach Machines To Think," *The Atlantic*, November 2013.

Chapter 12: Abductive Inference

1. Charles Sanders Peirce, "Guessing," *The Hound and Horn*, 2:267–282, 271.

2. Ibid., 271.

3. Ibid., 272.

4. Ibid., 277.

5. Gary Marcus and Ernest Davis, *Rebooting AI: Building Artificial Intelligence We Can Trust* (New York: Pantheon, 2019), 145–146.

6. Ibid., 146.

7. *Collected Papers of Charles Sanders Peirce*, eds. Charles Hartshorne and Paul Weiss, vols. 1–6, (Cambridge, MA: Harvard University Press, 1931–5), 5.189.

8. Charles Sanders Peirce. Charles Sanders Peirce Papers. Houghton Library, Harvard University, ms 692.

9. Melanie Mitchell, *Artificial Intelligence: A Guide for Thinking Humans* (New York: Farrar, Straus, and Giroux, 2019).

10. Ibid.

11. Peirce Papers, ms 692.

12. Collected Papers, 5.171.

13. Work on abduction experienced a partial revival in the 2010s, notably with attempts to cast it as a type of probabilistic or Bayesian inference. The approaches are worth the attention of specialists and non-specialists alike, if only because they help us further appreciate the true mystery of abduction viewed correctly as a guess or hypothesis that explains an event or observation in everyday (and scientific) thinking. The Bayesian systems invariably

make simplifying assumptions that on the one hand consider only a small subset of known possible hypotheses (sometimes as few as two), and on the other hand seek inferences that have a high probability given the example or case in question. The systems, while they are interesting extensions and explorations of probabilistic reasoning, are, again, abduction in name only. For a good summary of research results on the topic see Ray Mooney's work at the University of Texas at Austin, http://www.cs.utexas.edu/~ml/publications/area/65/abduction.

14. Hector Levesque, *Common Sense, the Turing Test, and the Quest for Real AI* (Cambridge, MA: MIT Press, 2017).

15. Hector Levesque, "On Our Best Behavior," *Artificial Intelligence* 212, no. 1 (2014): 27–35.

16. Large-scale projects scouring the web to cull facts and knowledge from it have also been disappointing. For discussion of Tom Mitchell's NELL (Never-Ending Language Learner), see Marcus and Davis, *Rebooting AI,* 150–151.

17. John Haugeland pointed out the problem of organizing knowledge for inference back in 1979: "the concept for 'monkey' would include not only that they are primates of a certain sort [taxonomic information], but also a lot of 'incidental' information like where they come from, what they eat, how organ grinders used them, and what the big one at the zoo throws at spectators." Haugeland's question was how to use "typical" information about the monkey concept when none of the encyclopedic information is relevant. This is still the question today. See John Haugeland, "Understanding Natural Language," *The Journal of Philosophy* 76, no. 11 (1979): 623.

18. R. C. Schank, *Conceptual Information Processing* (New York: Elsevier, 1975).

19. Peirce, "Guessing," 269.

20. Ibid., 269.

21. Ibid., 269.

22. Daniel Kahneman, *Thinking, Fast and Slow* (New York: Farrar, Straus, and Giroux, 2013).

23. Umberto Eco and Thomas A. Sebeok, eds., *Dupin, Holmes, Peirce: The Sign of Three* (Bloomington, IN: Indiana University Press, 1983).

Chapter 13: Inference and Language I

1. Andrew Griffin, "Turing Test Breakthrough as Super-Computer Becomes First to Convince Us It's Human," *Independent,* June 8, 2014.

2. See "Computer AI Passes Turing Test in 'World First,'" *BBC News*, June 9, 2014. The article from *Time* is no longer retrievable. See also Pranav Dixit, "A Computer Program Has Passed the Turing Test for the First Time," *Gizmodo*, June 8, 2014.

3. Gary Marcus, "What Comes After The Turing Test?" *New Yorker*, June 9, 2014.

4. Adam Mann, "That Computer Actually Got an F on the Turing Test," *Wired*, June 9, 2014.

5. We could convert Siri or Cortana into a competitor for the Loebner Prize by adding some code like doBabble() or doComplain(), where the arguments are questions or commands from the user. The system would then reliably carp about human requests—say, by always asking *why?* and then going on like an unwanted guest about how it was tired of always answering that or demanding to talk about something else. Siri might then seem generally mindful and intelligent, while also becoming completely useless. Siri will become more useful precisely to the degree that it actually understands human language, which is one reason why the Turing test remains unsolved, and perhaps also why AI scientists seem so keen to dismiss it.

6. Stuart Russell, *Human Compatible: Artificial Intelligence and the Problem of Control* (New York: Viking, 2019).

7. Martin Ford, *Architects of Intelligence: The Truth about AI from the People Building It* (Birmingham, UK: Packt Publishing, 2018).

8. Gary Marcus and Ernest Davis, *Rebooting AI: Building Artificial Intelligence We Can Trust* (New York: Pantheon Books, 2019), 6–7.

9. See Gary Marcus, "Why Can't My Computer Understand Me?" *New Yorker*, August 14, 2013. Levesque's paper, referenced in it, can \ be found here: https://www.cs.toronto.edu/~hector/Papers/ijcai-13-paper.pdf. Marcus's treatment of Levesque's work is insightful and worth reading. Levesque's paper for the IJCAI is readable by non-specialists and provides a great summary of the problem of big data for language understanding.

10. The last sponsored Winograd Schema Challenge was held by the International Joint Conferences on Artificial Intelligence in 2016. The winning system achieved 58.3 percent accuracy on the dataset, which did not qualify it for the prize. The sponsor has since declined to underwrite the $25,000 prize.

11. Hector Levesque, "On Our Best Behavior," lecture, International Joint Conference on Artificial Intelligence, Beijing, China, 2013, 4, https://www.cs.toronto.edu/~hector/Papers/ijcai-13-paper.pdf.

Chapter 14: Inference and Language II

1. Pragmatic analysis is the Holy Grail of Big Tech. Google would love to place ads for cheap airfare to Mazatlán, or maybe cheap umbrellas, on a web page near a comment that "The weather here is awesome!!!" made in Seattle in December (it's likely raining and cold). SPF 50 sunscreen and umbrella drinks are more likely ad placements, unfortunately. Sarcasm detection is (mostly) easy for people, but outside the realm of current AI.

2. A. M. Turing, "Computing Machinery and Intelligence," *Mind* 59, no. 236 (October 1950): 446.

3. This example also highlights contextual usage of the preposition *on*. We can be *on* a ship, but only *in* a car. Likewise, most people prefer *on* the train rather than *in*. Airplanes are tricky: I'm on the flight, but you're in the plane. But these examples can be captured by patterns of usage, since they are typical expressions found in texts. Problem cases involve atypical usage: *I'm on the house*—literally, I'm sitting on (top of) the house, which is an example of elision, leaving things out because they are understood in context.

4. A Google Brain team achieved 61.5 percent accuracy on a difficult Winograd schema test known as WSC-273. The state-of-the-art system managed a 10 percent increase in accuracy over random guessing, which shows more the insolubility of the test using data methods, rather than a legitimate advance. Note also that the team had access to the dataset used for training and testing, and extensively analyzed the questions—hardly an "out of the blue" test of language ability as any human would, presumably, be happy to perform.

5. Paul Grice, *Studies in the Way of Words* (Cambridge, MA: Harvard University Press, 1991).

6. Levesque, "On Our Best Behavior," 1.

7. In Winograd schema tests, some of the questions missed by AI systems tend to look similar and no more difficult than hits (correct answers). This scenario is bad, too, because then we have no way of accounting for the errors. A system simply gets some questions wrong that look for all intents and purposes exactly like the ones it gets right, the only difference being that, like picking at random, the system doesn't understand anything in the first place. Recent discussions about trusting modern AI systems when we don't understand why they go wrong (or are right) get to the heart of this issue, which is increasingly troubling as data-driven AI becomes more pervasive.

8. D. A. Ferrucci, "Introduction to 'This is Watson,'" *IBM Journal of Research and Development* 56, no. 3.4 (April 3, 2012), 1:1–1:15.

9. See Marvin Minsky, *The Society of Mind* (New York: Simon & Schuster, 1988).

10. J. Fan, A. Kalyanpur, D. C. Gondek, D. A. Ferrucci, "Automatic Knowledge Extraction from Documents," *IBM Journal of Research and Development* 56, no. 3.4 (2012), 5:1–5:10.

11. Marcus and Davis, *Rebooting AI*, 14.

12. Marcus and Davis discuss the differences between factoid and "common sense" questions in their section on language understanding and AI. See, for instance, ibid, 74–76.

13. Quoted in ibid., 68.

14. Ibid., 69.

15. Ibid., 27.

16. Charles Sanders Peirce, "Logical Machines," *American Journal of Psychology* 1, no. 1 (1887): 165.

17. Ibid., 168.

18. Ibid., 169.

19. Ibid., 169.

Chapter 15: Myths and Heroes

1. Jaron Lanier, *You Are Not a Gadget: A Manifesto* (New York: Alfred A. Knopf, 2010), 2.

2. Clay Shirky, *Cognitive Surplus: Creativity and Generosity in a Connected Age* (New York: Penguin Press, 2010).

3. Yochai Benkler, *The Wealth of Networks: How Social Production Transforms Markets and Freedom* (New Haven, CT: Yale University Press, 2007).

4. Ibid., *Epigraph*.

5. Clay Shirky, *Here Comes Everybody: The Power of Organizing Without Organizations* (New York: Penguin Books, 2009).

6. Lanier, *You Are Not a Gadget*, 1.

7. James Surowiecki, *The Wisdom of Crowds* (New York: Anchor, 2005).

Chapter 16: AI Mythology Invades Neuroscience

1. Sean Hill, "Simulating the Brain," in Gary Marcus and Jeremy Freeman, eds., *The Future of the Brain: Essays by the World's Leading Neuroscientists* (Princeton, NJ: Princeton University Press, 2015), 123–124.

2. See Henry Markram, "Seven Challenges for Neuroscience" *Functional Neurology* 28 (2013): 145–151.

3. Ed Yong, "The Human Brain Project Hasn't Lived Up to Its Promise," *The Atlantic*, July 22, 2019.

4. The Kavli Foundation, "The BRAIN Initiative: Surviving the Data Deluge," https://www.kavlifoundation.org/science-spotlights/brain-initiative-surviving-data-deluge#.XgVezkdKhdg.

5. Amye Kenall, "Building the Brain: The Human Brain Project and the New Supercomputer," *BioMed Central*, July 8, 2014, http://blogs.biomedcentral.com/bmcblog/2014/07/08/building-the-brain-the-human-brain-project-and-the-new-supercomputer/.

6. Rebecca Golden, "Mind-Boggling Numbers: Genetic Expression in the Human Brain," *Science 2.0*, April 15, 2013, https://www.science20.com/rebecca_goldin/mindboggling_numbers_genetic_expression_human_brain-109345.

7. Yves Frégnac and Gilles Laurent, "Neuroscience: Where Is the Brain in the Human Brain Project?," *Nature*, September 3, 2014.

8. Ibid.

9. Ibid.

10. Eric Kandel, Henry Markram, Paul M. Matthews, Rafael Yuste, and Christof Koch, "Neuroscience Thinks Big (and Collaboratively)," *Nature Reviews Neuroscience* 14, no. 9 (2013): 659.

11. Lee Gomes, "Machine-Learning Maestro Michael Jordan on the Delusions of Big Data and Other Huge Engineering Efforts," *IEEE Spectrum*, October 20, 2014, https://spectrum.ieee.org/robotics/artificial-intelligence/machinelearning-maestro-michael-jordan-on-the-delusions-of-big-data-and-other-huge-engineering-efforts.

12. Ibid.

13. Gary Marcus and Ernest Davis, "Eight (No Nine!) Problems with Big Data," *New York Times*, April 6, 2014.

14. Nate Silver, *The Signal and the Noise: Why So Many Predictions Fail—But Some Don't* (New York: Penguin Books, 2015).

Chapter 17: Neocortical Theories of Human Intelligence

1. Jeff Hawkins, *On Intelligence: How a New Understanding of the Brain Will Lead to the Creation of Truly Intelligent Machines* (New York: St. Martin's Griffin, 2005).

2. Ray Kurzweil, *How to Create a Mind: The Secret of Human Thought Revealed* (New York: Penguin Books, 2013), 35.

3. Gary Marcus, "Ray Kurzweil's Dubious New Theory of Mind," *New Yorker,* November 15, 2012.

4. Ibid.

5. For a general description, see Ferris Jabr, "Memory May be Built with Standard Building Blocks," *New Scientist,* March 17, 2011.

6. Henry Markram and Rodrigo Perin, "Innate Neural Assemblies for Lego Memory," *Frontiers in Neural Circuits* 5 (2011): 6.

7. Eric Kandel, Henry Markram, Paul M. Matthews, Rafael Yuste, and Christof Koch, "Neuroscience Thinks Big (and Collaboratively)," *Nature Reviews Neuroscience* 14, no. 9 (2013): 659.

8. Yves Frégnac and Gilles Laurent, "Neuroscience: Where Is the Brain in the Human Brain Project?" *Nature News* 513, no. 7516 (2014): 27

9. "Open message to the European Commission concerning the Human Brain Project." See https://neurofuture.eu/.

10. Eliza Shapiro, "Obama Launches BRAIN Initiative to Map the Human Brain," *Daily Beast,* April 3, 2013. Updated July 11, 2017.

11. See Stefan Thiel, "Why the Human Brain Project Went Wrong—and How to Fix It," *Scientific American,* October 1, 2015. Also Ed Yong, "The Human Brain Project Hasn't Lived Up to Its Promise," *The Atlantic,* July 22, 2019.

12. Yong, "The Human Brain Project Hasn't Lived Up."

Chapter 18: The End of Science?

1. Peter Thiel, interviewed by Eric Weinstein on "The Portal" podcast, Episode #001: "An Era of Stagnation & Universal Institutional Failure," July 19, 2019, https://www.youtube.com/watch?v=nM9foW2KD5s&t=1216s.

2. Ibid.

3. Norbert Wiener, *Invention: The Care and Feeding of Ideas* (Cambridge, MA: MIT Press, 1994).

4. Ibid., 89.

5. Ibid., 96.

6. Ibid., 96.

7. Ibid., 87.

8. Ibid., 87.

9. John Horgan, *The End of Science: Facing the Limits of Knowledge in the Twilight of the Scientific Age* (Boston: Addison-Wesley, 1996).

ACKNOWLEDGMENTS

Many people were involved both formally and informally in the creation of this book. First, on the Harvard University Press team, I would like to thank Jeff Dean, now at Hackett Publishing, for his early support of the project while an editor at HUP. Jeff helpfully pushed back on my thinking in the earliest attempts at putting pen to paper, and I am grateful for his patience in getting past the first bumps. James Brandt, Joy de Menil, and Julia Kirby picked up where Jeff left off and made the text better in their own ways. Graciela Galup contributed elegant design and Colleen Lanick and her publicity team found creative ways to get the word out.

Many other people played significant and much appreciated roles in helping see the book through. I was overseas for part of the writing of the manuscript, and events in the States made things difficult at times. I am grateful to Bernard Fickser for providing funding through a generous deal on a jointly founded startup, as well as his encouragement in a number of ways too various to mention. Todd Hughes was an early reader of the manuscript and made very helpful suggestions. My agent Roger Freet, now at Folio Literary Agency, was tireless in explaining to me many aspects of book writing and publishing, and his assistance on the proposal was much appreciated.

I also would like to thank Anna Samsonova for putting up with me through those long days in Europe, and for helping me through the interminable tangles of learning Russian while trying to formulate my thoughts in English. I would also like to thank my writing friends in Seattle (you know who you are) who made invaluable and still somewhat unacknowledged contributions.

Finally, to my friend John Horgan, the great science writer and an inspiration to so many, thanks so much for your early belief in the book, and for your consistent encouragement.

INDEX